LOCAL LEVEL PLANNING AND RURAL DEVELOPMENT

LOCAL LEVEL PLANNING AND RURAL DEVELOPMENT

Alternative Strategies

United Nations

Asian and Pacific Development Institute

Bangkok

CONCEPT PUBLISHING COMPANY, NEW DELHI-110059

"The views expressed in the signed papers do not necessarily reflect those of the United Nations"

ISBN 81-7022-102-1

First Published 1980
Reprinted 1986, 2006

Published and Printed by
Ashok Kumar Mittal
Concept Publishing Company
A/15-16, Commercial Block, Mohan Garden,
New Delhi-110059 (India)

Phones : 25351460, 25351794
Fax : +91-11-25357103
Email : publishing@conceptpub.com

FOREWORD

GIVEN the current misgivings about the appropriateness of existing top-down strategies of development planning, the publication of this book is a very timely one. It addresses in a very direct way the issues involved in people-oriented development planning, explores the conceptual and operational problems of an alternative development strategy with focus on people's participation and local level planning, and reports very usefully on the current experiences with this strategy in various countries in the Asia-Pacific region.

The papers and reports contained in this book are the result of two expert consultative meetings held by the Economic and Social Commission for Asia and the Pacific (ESCAP) and the United Nations Asian and Pacific Development Institute (UNAPDI) held in Bangkok in October-November 1978. The two meetings were interlinked and jointly planned by ESCAP and UNAPDI. The first meeting discussed alternative development strategies for the predominantly agrarian economies of the Asian and the Pacific region and provided the broad perspective and guidelines for the second meeting which specifically focused on local level planning for integrated rural development. These meetings were held because it was felt necessary to evaluate the effectiveness of our actions in alleviating the abject poverty suffered by millions of the poor and the underprivileged who live below subsistence level and are without the means to meet even their basic minimum needs.

The alleviation of this poverty is now widely accepted at both national and international levels as a task of first priority. In the work programme of ESCAP, and also of the associated institutions, such as the UNAPDI, "eradication of poverty, unemployment and social inequality", to quote the central theme of UNAPDI, has been recognized as the main preoccupation. ESCAP has worked closely in partnership with the specialized agencies and international organizations within the United Nations system in drawing up an interagency coordinated plan of action for integrated rural development in the ESCAP region in response to Commission Resolution 161 (XXXII), adopted at its thirty-second session in 1976.

The Commission considered the plan at its thirty-third

session in 1977, and directed the Executive Secretary to implement it as a matter of highest priority in consultation with member Governments and in partnership with concerned specialized agencies and international organizations.

The Interagency Committee on Rural Development for the ESCAP Region, composed of the heads of the regional offices of the collaborating agencies and organizations located in Bangkok, has been overseeing the implementation of the plan with the assistance of an Interagency Task Force. Both the Committee and the Task Force are serviced by an IRD Unit located in the Executive Secretary's Office. It is gratifying to note that the mechanism set up to promote interagency consultation and a co-ordinated approach to rural development in the participating countries has functioned satisfactorily and effectively during the period since its inception.

While considerable efforts have been made and are increasingly being made in rural development at both national and international levels, there is a need to re-examine the concepts, strategies and programmes with reference to the objectives sought, namely, self-sustained growth with equity, greater employment and improvement in the quality of life of the poor with their full participation as equal partners in development.

This is exactly what this book is about. The contributors to this volume are men rich in knowledge and experience with the trials and tribulations of development planning and they have here laid out the elements of a viable alternative strategy for rural development. For this reason this book should be of value not only to the general reader and the student of rural development but also to the planner who wants a quick review of what is going on in the ESCAP region with regard to development.

It is only left now for me to hope that the ideas included in this volume will bear fruit in the development decade of the eighties.

J.B.P. Maramis
Executive Secretary,
ESCAP
Bangkok
March 1980

CONTENTS

PART II

LOCAL LEVEL PLANNING

LIST OF CONTRIBUTORS

G.C. Clarke,

Leader, Small Farmers Development Team, FAO Regional Office for Asia and the Far East, Bangkok, Thailand

Kurt Dopfer,

Professor of Economics, Swiss National Fund for Scientific Research, Berne, Switzerland

Benito F. Estacio Jr.,

B. Morcilla St. Peters, Metro Manila, Philippines

*L.S. Frederics,**

Deputy Dean, Faculty of Economics and Administrations, University of Malaya, Kuala Lumpur, Malaysia

Shoaib Sultan Khan,

Islamabad

A.Z.M. Obaidullah Khan,

Secretary, Ministry of Agriculture and Forests, Government of the People's Republic of Bangladesh

Ram C. Malhotra,

Deputy Director, United Nations Asian and Pacific Development Institute (UNAPDI) Bangkok, Thailand

*K.P.A. Menon,**

Additional Secretary, Department of Rural Development Ministry of Agriculture and Irrigation, New Delhi, India

National Institute of Public Administration (LAN)

Jakarta, Indonesia

Salima Omer,

Associate Professor, University of Nebraska, USA, Currently, Senior Regional Expert, ESCAP, Bangkok, Thailand

*K.P.G.M. Perera,**

Director, Regional Development Division, Ministry of Plan Implementation, Colombo, Sri Lanka

*Bihari K. Shrestha,**

Reader in Anthropology, Centre for Nepal and Asian Studies, Tribhuvan University, Kathmandu, Nepal

*Vithya Siripongse,**

Director, Regional Planning Division, National Economic and Social Development Board, Bangkok, Thailand

Vinyu Vichit Vadakan,

Director, United Nations Asian and Pacific Development Institute (UNAPDI) Bangkok, Thailand

*B.W.M. Van Dusseldorp,**

Associate Professor, Agricultural University, Wageningen, Netherland

Ram P. Yadav,

Executive Director, Agricultural Projects Services Centre (APROSC) Kathmandu, Nepal

* The titles of these authors are as of November 1978.

PART I

ALTERNATIVE DEVELOPMENT STRATEGIES

INTRODUCTION

ONE of the major objectives of the United Nations Asian and Pacific Development Institute (UNAPDI), as a regional institution engaged in training and research for development, has been "to assist countries in the ESCAP region in the formulation of appropriate concepts, strategies and techniques for development, relevant to conditions and problems in the Asian and the Pacific region". UNAPDI has also adopted since 1975-76 "eradication of poverty, unemployment and social inequality" as the central theme of its work programme. These objectives have been formulated directly in response to an ever growing search in almost all the countries in the region for appropriate and viable "alternative strategies for development" largely motivated by the concern that the development strategies pursued during the past quarter of a century or so have failed to benefit demonstrably the majority of the people or to involve them in development.

UNAPDI has been engaged since late 1974 in an attempt to develop a conceptual framework for rural development at both the macro and the micro levels. UNAPDI also felt it necessary to place a rural development programme in the broader perspective of an "alternative" strategy of development and also to examine closely the operational and institutional aspect of both an alternative development strategy and an IRD programme including local level planning and development. Accordingly, UNAPDI has included in its annual work programmes research projects on "alternative strategies for development", "evaluation of integrated rural development programmes" and "local level planning and development" to be conducted at two levels namely, (a) conceptual, with work to be done mostly at the regional or inter-country level and based on the experience of the countries in the region or elsewhere, and hence in collaboration with expertise mainly drawn from the region, and (b) operational or country-specific, with work to be done at the country level in interested countries and in collaboration mainly with national expertise drawn from all available sources, viz. public and private sectors and academic institutions etc.

In order to seek further guidance for work at both levels, UNAPDI held an expert Consultative Meeting on Alternative Strategies for Development Focusing on Local-Level Planning and Development in Bangkok from 21 October to 4 November 1978. This Meeting was held in collaboration with ESCAP as the first in a series of two meetings. The second was held by ESCAP in collaboration with UNAPDI on "Local Level Planning for Integrated Rural Development" in Bangkok from 6 to 10 November 1978. The collaboration of the Interagency Task Force on Integrated Rural Development was sought in undertaking both these activities.

The objectives of the first meeting, broadly stated, were (a) to systematically identify elements and clarify concepts in outlining alternative or new strategies or approaches to development, and (b) to delineate the role and scope of local-level planning in the context of the new approaches to development.

Any strategy of development being country-specific, an alternative strategy or strategies, it is felt, may be described as new approaches only in terms of the attributes of major objectives of such approaches, these being relieving abject poverty by providing employment to the rural and urban poor, provision to meet socially acceptable basic minimum needs of all, higher productivity, greater equity between geographical areas and among classes of people, decentralization and people's involvement in decision-making, emphasis on collective and national self-reliance and environmental balance and enrichment.

The task of clarification of concepts and delineation of common elements and issues involved in outlining alternative development strategies for the predominantly rural economies of the Asia and Pacific region was undertaken on the basis of an examination of innovative measures and an evaluation of new approaches to development in various countries of the region, for achieving the commonly accepted major development objectives broadly stated above.

The expert Consultative Meeting was thus expected to focus mainly on operational aspects of "alternative development strategies", in order to develop a set of guidelines for further work both at UNAPDI and at the country level by selected national institutions or multi-disciplinary research teams. The task was to systematically work out the steps and appropriate policy and other measures, including institutional mechanisms for the formulation, implementation, monitoring and evaluation of country-specific alternative development strategies, including rural development programmes and local-level planning and development. A detailed examination of local-level planning and development, based on a review of experience in selected countries, was made at the second meeting, already referred to, held by ESCAP in collaboration with UNAPDI at Bangkok from 6 to 10 November 1978.

The Consultative Meeting made a number of suggestions for further research and experimentation in the area of local level planning and development. Accordingly UNAPDI initiated in 1979, with UNDP programme support and in collaboration with national governments and national research institutions, a research project on Local Level Planning and Development in five countries in Asia. A note on the research project is given in the appendix.

The task of formulating endogenous development strategies is urgent in the ESCAP region. The task however is a stupendous one. One uneasy impression that we have carried from the UNAPDI Consultative Meeting on "Alternative Strategies" is how little we seem to know about the task before us and its numerous facets. Even after nearly three decades of experience with planning and development in the developing world, it appears that we are still scratching the surface of the basic goal of human development. We sincerely hope this publication, originally brought out in February 1979 in a limited number of copies in mimeographed form, can now reach a wider audience of policy makers and planners and make a modest but useful contribution to a better designing of appropriate development strategies in the countries in the ESCAP region.

UNAPDI will be terminating at the end of June 1980 after seventeen years of its useful existence as a regional training and research institution. We sincerely hope that the research work initiated by it in the area of local level development will be continued by the participating countries and UNAPDI's successor institution, namely, the UN Asian and Pacific Development Centre (APDC) to be located in Kuala Lumpur beginning on 1 July 1980.

Ram C. Malhotra, Deputy Director of the Institute, acted as the coordinator of the UNAPDI Consultative Meeting and also contributed the principal background paper for the meeting. He has supervised the Institute's programmes in rural development and is also the coordinator of the UNAPDI research project on Local Level Planning and Development, already referred to. He has edited the papers in this publication with the assistance of Dr. T.P. Jagdeo, Consultant with UNAPDI. I should like to record our thanks to both of them and also to all other colleagues and staff members of the Institute who have rendered valuable assistance towards this publication.

Vinyu Vichit Vadakan
Director
United Nations Asian and
Pacific Development Institute
(UNAPDI)
Bangkok
10 March 1980

ONE

AN ALTERNATIVE STRATEGY FOR SELF-SUSTAINING DEVELOPMENT WITH FOCUS ON PARTICIPATION BY THE POOR AT THE LOCAL LEVEL

Ram C. Malhotra

Introduction

A GOOD DEAL of literature has appeared in recent years, and is continuing to appear, on rural development, people's participation, basic minimum needs and target-group-oriented approaches to development and related topics. A review of this literature would suggest that a broad consensus has been emerging on what went wrong in the past and on what new approaches to adopt in the future. However, there is unfortunately very little written about *how* the new approaches should be implemented. The purpose of this paper is to focus on the latter, on the *operational* aspects of a viable development strategy designed to accomplish the now widely accepted objectives of integrated development. The paper will therefore start from where "what-should-be-done" analyses and conceptual frameworks[1] have left off, and attempt to delineate the steps necessary for designing an operational strategy for growth-with-equity and target-group-oriented micro-level development with people's participation, particularly the poorest, on a self-sustaining or self-reliant basis, in the context of the predominantly agrarian economies of Asia and the Pacific region.

This paper however has a modest scope. It does not pretend to discuss or even raise all the issues and dimensions related to an alternative development strategy in both the national and international contexts. These important issues are not intended to be ignored and are implicit in the paper. But its focus is mainly on the national context, and further limited to the issue of mobilisation of surplus human labour, the single resource available in abundance in agrarian economies of the Third World, for providing gainful employment and meeting the basic needs of the masses, in an equitable and self-sustaining manner.

The paper is divided into three parts. In the first part, the commonly accepted objectives are recapitulated and restated for purposes of arriving at a common understanding of these objectives, and also with a view to providing a set of criteria for evaluating the appropriateness of a development strategy, and the institutional and procedural mechanisms and models for implementing the strategy, as suggested in the latter parts of the paper for the achievement of the stated objectives. Any strategy of development must in the first analysis be judged in terms of the objectives. This is what, in the new discipline of development administration, is known as "Management by Objectives". Management or administration of development, as used here, must be understood in its broadest sense, that is to say, management or mobilization of all available resources—natural, human, cultural, social, institutional, political—and putting them to optimum use for the achievement of the stated goals of development. Hence it encompasses planning, implementation, monitoring and evaluation; that is to say, all the processes of development.

Objectives of Development

The now commonly accepted objectives of a target-group-oriented anti-poverty strategy of development with people's effective participation are restated as follows :

1. *Higher productivity and increase in production or growth in GNP*. Economic growth, although much maligned in recent years, is still necessary; but it should not be the only goal and must be seen as one of the components of an "integrated" or "unified" approach to development. The other equally important accompanying or inseparably interlinked set of goals are as follows.
2. *Distributive justice or equity* in sharing, or distribution of benefits of development, or what may also be referred to as the 'people's participation in benefits';

equity must be achieved in meeting *basic minimum needs* for the underprivileged section of the population or in terms of the "elimination of abject poverty". "Basic minimum needs" include primary economic and social needs such as food, shelter, clothing, basic education and health and other public services such as clean drinking water, and finally, production inputs such as farm inputs, etc. The concept of "basic minimum needs" is not only physiological but also a moving or dynamic concept and relates to the socially acceptable minimum standard of living commensurate with the stage of development in a country at a given time.

3. *Employment* is both a means as well as an end in itself, in terms of the use of labour resources for production and for providing income to the people as well as dignity and meaning to the lives of individuals
4. *People's participation*, like employment, is also both a means and an end—but I am inclined to look upon it more as an end in itself—in social and political terms, as it is the *sine qua non* of development, particularly at the local level. And by people's participation, we mean willing and voluntary participation; it may be spontaneous or induced, but certainly not "coerced", for that is not participation. Just as implementation is the touchstone for planning, people's participation may be looked upon as the touchstone of "unified" or "integrated" approach to development.

   Corollaries of, or rather the prerequisites for, people's participation are: a democratic system which will allow the people, even the smallest and the poorest, to participate directly as well as through genuine representatives in the decision making process at all levels, and also that will provide for devolution of political power to the smallest village, and decentralization of administrative authority to the lowest level in the administrative hierarchy.

   These are means of such great significance and importance for people's participation that these may be looked upon as ends in themselves. People's participation, in turn, is also expected to lead to the inculcation or development of democratic values, and thereby to strengthening democracy, which is an end in itself.

5. Next, *self-reliance* may be looked upon as a vital component of the integrated goals of development. Self-reliance, to my mind, is the will to make use of all available domestic resources and to put reliance on one's own resources for self-sustaining development. External aid should be made use of as supplementary to domestic resources and not as the primary or the major resource, as has been the case with many a development plan in many a country over the last so many years. Self-reliance again must be looked upon as a goal worthy of emulation not only at the national level but at all levels—local as well as intermediate.
6. Finally, maintenance of an *ecological balance* between man and nature, that is to say ecologically sound use and management of finite and renewable natural resources such as soil, water, air and forests, etc. must be included not only as a means but as a desirable objective of a well integrated development strategy.

We may agree on these six, or if we include devolution of political power and decentralization of administrative authority and/or inculcation of democratic values as additional goals, seven or eight major goals*. Additional goals based on national priorities may also be added to the list. Two points need to be stressed.

* The hierarchy of goals of development may be shown in the form of a pyramid as follows, wherein at the base are the economic and social "basic minimum needs" for subsistence whose fulfilment leads to a higher set of socio-political needs, and ultimately to the goal of the full flowering of the human personality or "total development", and the release of creative energies of every individual.

First, these goals are not only *inter-dependent and inter-linked* but must be looked upon as a *single package*. Although there are 'contradictions' and possible trade-offs among them, we should not think in terms of completely neglecting or eliminating any one of these goals. For we now know, after twenty-five years or so of bitter experience, the consequences of trade-off practised between growth and equity. The two must be inalienably interlinked. Similarly, the goals of employment, people's participation and self reliance are closely interlinked and so on.

In this context, an *integrated approach to development*, I should like to submit, lies more *in terms of objectives* than in terms of sectoral integration. Any programme or project for sectoral development may be implemented in an integrated manner provided the instruments for implementation, i.e.

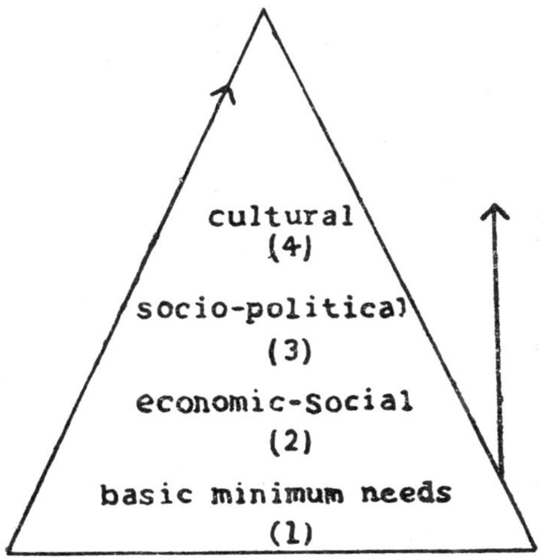

(4) full development of human potentials and creativity;
(3) equity or distributive justice, social equality, redistribution of assets, democracy;
(2) higher growth with greater equity, mass consciousness;
(1) eradication of abject poverty and unemployment—access to minimum income and public services through employment with people's participation.

The level of development is subject to the degree of "mass consciousness", that is to say the higher the level of mass consciousness, the higher will be the level of development achieved.

institutions and procedures and the approach is based on integrated objectives as stated above. This may well be taken as an issue for discussion. In operational terms this would mean that an integrated programme or project at the village level need not be a multi-dimensional complex affair but can be a very simple activity.*

Second, as it was already stated in the beginning, it is important to be clear about these objectives in order to be able to devise an appropriate strategy and mechanisms, or administrative structures and procedures including devices for monitoring and evaluation, and also to be able to monitor and evaluate, that the management devices themselves are appropriate and are on course and have not been devaluated or derogated by vested interests, as very often happens. This to my mind is the purpose of management by objectives i.e. the management of development programmes and projects for stated objectives of total development and the evaluation of appropriateness of every means adopted as well as the performance or efficacy of management, in the broadest sense as defined earlier, against the criteria or yardsticks provided by the objectives.

Hence, we turn next to an examination of the implications or underlying assumptions of those objectives.

Underlying Assumptions and Implications of the Objectives

Mixed-Economies

The objectives as stated above, and any development strategy based on these objectives, are designed for the type of "mixed economies" that are prevalent today in a majority of the countries of the developing Third World. Admittedly there is a wide variety of "mixed-economies" in existence in the Third World and it will be difficult to classify them. But, as distinct from predominantly capitalistic or purely socialist economies, these mixed economies allow for co-existence of both the public and

* This was a point stressed by several participants in two subregional seminars (Manila/Los Banos, 9-13 January 1978 and Dacca/Comilla, 23-27 January 1978) organized by the Asian Development Bank on the Second Agriculture Survey. It is essential to avoid over-elaboration of plans in order not to make them complicated for participating beneficiaries, such as the small farmers.

the private sectors. The public sector's role is understood to be mainly focused on providing infrastructures like roads and other public works, power and irrigation projects, public utilities for mass transportation and communication, social services like health and education and certain basic industries, the latter mainly through public corporations. The private sector is given freedom, subject to regulatory controls as well as incentives, for production and distribution of goods and services in almost all sectors of the economy namely agriculture, industry, tourism, social services, and so on.

People's Sector

In addition to the well-known two organized sectors, viz. public and private, a third sector, what I would like to call "the people's sector", may be identified. It is the vast decentralized and unorganized sector of the majority of the people mostly living in rural areas in the predominantly rural economies of the Third World. It is with this sector that the development strategies discussed in this paper are concerned. An implicit assumption made in the paper is that the public and the private sectors will continue to perform their respective roles in national development as normally understood; and also that the public sector will assume a catalytic and promotional role for the growth and sustenance of the people's sector.

Also implied in the above assumption is that the existing socio-economic political structure should be accepted as a given parameter and that the existing power structure cannot be wished away. Nor should one expect the existing power base to work against its own interests for the betterment of the lot of the poor. Hence, any development strategy that aims at ameliorating the lot of the poor and the underprivileged in a "mixed-economy" will have to be such as would, at least in its initial stages, fit the norms of the society and be acceptable to the existing power structure. But it should have built-in mechanisms for mobilizing the poor and raising their consciousness regarding their rights and privileges on the one hand*, and

* For one lesson that has emerged invariably from all analysis of development strategies practised during the past quarter of a century is that the poor must be organized before they can claim their legitimate share of the fruits of development.

for concurrently bringing about an awareness among the rich and the privileged of the need to work towards a more egalitarian society in their own "enlightened self interest". Development may therefore be seen as a process of education for all sections of society in a mixed-economy for harmonious growth with equity.

"Equity" in Basic Needs

This would suggest that the goal of equity cannot be reached overnight, and that the achievement of the goal will be commensurate with the strengthening of the political power base of the poor, which in turn would depend on the degree of their own awareness about their rights and privileges and their mobilization as a socio-political force in a community.*

The "basic needs" approach that is incorporated in the goal of equity, as discussed earlier, also implies that "equity" is a goal that can be reached only gradually or by stages. What is aimed at in the initial stages, and perhaps also to be accomplished over a number of years, is eradication of abject poverty or guarantee of equity only at the socially acceptable minimum or subsistence level Inequity above the "basic minimum needs" line will continue until this line approaches the level of living of the richest in a community. This may not be possible for a long time to come or until there is complete redistribution of all wealth and all the means of production. Hence, given the pattern of socio-political power structures in the mixed economies of today, to regard "redistribution of wealth" as a precondition for alternative development may not be realistic.

Preconditions for Development

Looked at from the same angle, any other "preconditions" for an alternative development strategy, such as land reform, depending on the socio-political structure or the existing level of mass consciousness in a given country, may as well be looked upon as intermediate goals in a hierarchy of ends and means

* This may suggest "gradualism" but there is no short-cut to development, not even through a revolution, for revolutions also take a long time in the making and require a committed leadership and the building of a cadre of workers, if the Chinese experience is any guide to a successful revolution in recent years.

wherein the lower ends become means to higher goals. "A pre-condition for the success of a land reform programme is the organization of beneficiaries."[2]

There is however, one precondition for any development that cannot be dispensed with, and that is *committed leadership* or *political will* on the part of the highest leadership in a country to work for the achievement of the objectives as stated above. There is no substitute for committed leadership at the top. The commitment however should not be to anything more radical than the limited package of objectives as stated above. The task of the leadership at the top and at all levels would then be to design an appropriate development strategy for achieving the package of objectives, and to rigorously apply the criteria, provided by or derived from those objectives, to the choice of institutions and policy and other instruments for implementing the strategy or strategies, and also for monitoring and evaluating the results achieved. The leadership will haveto be always alert and also have a capacity to *think-through* both the short and long-term effects of alternative measures that may be suggested for achieving the integrated set of objectives.

This task is certainly not beyond the grasp of leadership in any developing country, provided that the leadership is honest and has the vision and statesmanship to be able to look beyond immediate personal and sectional gains to the long-term "enlightened self-interest" of all sections in the society as a whole.

II

Strategy of Development

One common characteristic of the developing countries of the Third World is that they are predominantly rural. Eighty per-cent or more of their populations live in the rural areas, and again almost 80 per cent of this rural population is engaged in agriculture. Also another characteristic they commonly share is that they are abundantly rich in at least one resource, namely human labour, which however is mostly unskilled and often illiterate. Most of this labour is derived from marginal farmers and landless farm labourers who are either underemployed on their farms, or seasonally unemployed when there is not much work to be done on the farms. It is estimated that agricultural work on the farms provides employment for a maximum of 180 days or 6 months in a year.*

This vast reservoir of surplus labour in a developing country is a valuable resource that could be put to use for development, particularly at community level, in the rural areas, or what I have described as the "people's sector". Unfortunately it is going to waste today. Labour if unused cannot be preserved, it being like time the most perishable commodity. Unused labour, like the unused vast water resources flowing down the Himalayas, is not only wasted like those rivers, but also, like the river waters that take with them the top soil of the Himalayan region causing permanent losses in soil erosion, cause permanent damage to the human will and spirit in a community through despondence and frustration born out of ever-increasing levels of abject poverty over the generations. It is the human spirit and the morale of the unemployed and the poor that gets slowly eroded, and yet it is the restoration and building of this human spirit and morale of the majority of the population, through their all round material and spiritual development by realising their creative energies, that development in the ultimate analysis is concerned with.

Surplus Labour

This strategy, alternative if you will, because the focus herein is on "surplus" labour and its capitalization rather than on "scarce" capital and its "efficient" use in terms of "capital output" ratios as has been the case with development strategies pursued so far, appears to be, to my mind, the only viable development strategy appropriate for the predominantly rural economies of the Third World. For this surplus labour resource, although unskilled in modern technology, comes with a lot of sound common sense and expertise in traditional

* For example, a recent survey revealed that in the hilly regions of Nepal, farming provides work for only 55 days in a year as compared to 180 days in the southern "terai" or the plains region of Nepal. A survey carried out by the National Planning Commission of Nepal in 1976 revealed that in 1964/65 nearly half of the labour force engaged in agriculture was unemployed. *Fifth Plan (1975-1980)*, Nepal, p. 46.

technology and is capable of building roads, small suspension and other bridges, minor irrigation works and channels, bunds for erosion control, reforestation, drinking water projects, school and hospital buildings in rural areas. It is also capable of being employed in a host of other developmental activities besides rural works, such as in small-scale industries, rural electrification, cooperative farming and marketing, etc. There is a crying need for all those development activities in the rural areas of the Third World. In the face of the proverbial paradox of poverty amidst the plenty of manpower and other natural resources, the task of development is to mobilize the labour resources for national development with particular reference to development at the local level. This would appear to be essentially a problem of resources management, defined in its broadest sense, within the framework of the given objectives.

Utilisation of surplus labour as a strategy for development, particularly rural works programmes, however, is not a new strategy. Indeed it has been employed since times immemorial; the Great Wall of China, the pyramids of Egypt and most of the monuments of ancient civilization all over the world are monuments to the use of labour. But the difference between the old strategy in feudal times and the new is that labour's "participation" in the former was most likely coerced. Can we achieve similar results by willing and voluntary participation of the masses including the poorest, on the basis of equality and partnership and without being exploited by the rich and the rulers as in feudal times?

The system of using "voluntary" labour for the construction of rural works either on the basis of contribution of free labour by the villagers, or under the various "food for works" schemes is also all too familiar to almost every rural society throughout the Third World. Based on experience with such schemes, there are two serious objections to such a practice. First, the rural works programme—the majority of these have been village roads—have too often been hastily conceived, ill-planned and executed without proper design, and most often have either remained incomplete or have been washed away by the rains or, if completed at all, have been short-lived for lack of regular maintenance. But the more serious objection from the points of both employment and equity is that it is almost invariably the poorest in the village, who are employed or underemployed, that are called upon to give free, or on the basis of partial payment, the only asset they possess, namely labour, in the name of the public good. Given the prevailing inequalities in the rural societies, such contribution by the poorest, without any or full payment is not only inequitable but also exploitative of the weaker sections. For, the rich and the privileged, who often stand to gain most by the construction of the rural works, are seldom called upon to make proportionate contributions either in cash or in labour to the rural works. Because of these reasons, particularly the latter, it it the thesis of this paper that the use of the so-called "voluntary" *free* or *partially paid* labour for rural development should be abandoned on the grounds of both equity and gainful employment for the poor.

This will invariably raise the problem of finding adequate financial resources for payment of wages to the unemployed at the going market rates or even at higher levels in order to attract them to jobs in the rural areas.

Also assuming that such large resources were available either to fully or even partially pay for mass employment schemes, there would remain the other problem of high levels of inflation, particularly in the prices of wage goods, caused by the additional purchasing power in the hands of the newly employed.

These are serious issues for which no viable answers appear to have been found as yet.

In the face of these insuperables, studies of investment of surplus labour as capital have often come to the conclusion that such investment is only possible under coercive systems such as the feudal regimes of the past or the socialist systems as prevalent in China.[3] By implication such studies imply that in democratic societies or market-oriented mixed-economies of the Third World there is no viable method to employ surplus labour into productive use.

This however is a thesis that needs to be examined more thoroughly as will be done in the balance of the paper. Hence first it is necessary to be clear about the nature of the issues involved.

It is not the use of labour for rural works or for any development programmes that is bad; on the other hand, perhaps the only way surplus labour can be capitalised is through the construction of rural works or its investment in other development projects for its use in productive ways.

It is the wasteful and haphazard use of such a valuable resource, and the exploitative and inequitable impact on the weakest sections of the rural society of the manner in which surplus labour is mobilized that are at fault.

The need is therefore to look closely at the institutions and methods for using surplus labour for purposes of rural development, and design them in such ways that would meet the criteria or the objectives discussed in the beginning of this paper.

Such a task is certainly not beyond human ingenuity. In other words, if we accept the objectives and the basic strategy discussed so far, then the major issue before the planners and administrators is to find ways and means and devise institutions and techniques whereby the single major resource available in abundance in the developing poor countries of the Third World can be put to use for meeting the basic needs of the masses, with particular focus on the poor and the underprivileged, in a self-sustaining or self-reliant manner. This with a view to providing them with access to employment and additional income and also greater access to public services, with dignity, self-respect and their willing and voluntary participation as equal partners in decision-making as well as in an equitable sharing of the benefits of development. It is this issue that this paper is primarily concerned with.

Basic Elements

The basic elements or requirements of a viable alternative model for use of surplus labour for development in the "people's sector", that is to say, at the local level would appear to be as follows :

Need to/for

(1) mobilise and channel vast resources of surplus labour into productive uses or creation of social overheads such as through rural works programmes, in an

equitable, non-exploitative manner and with willing and voluntary participation of the poor majority who are the unemployed and underemployed and have the labour to offer;

(2) pay wages at market rates or even higher rates in order to make employment in rural areas attractive to the unemployed or underemployed, in order both to stem their drift to urban centres and to provide them, if feasible, with proportionately higher incomes for reasons of equity; banish use of "free" labour, even though it were "voluntary", again for reasons of equity;

(3) provide the poor majority greater access to minimum public services such as health, education, drinking water, public transport and essential production inputs such as skills, tools, water for irrigation, new seeds, fertilisers and pesticides, etc;

(4) forestall inflation in prices of wage goods and essential public services due to increase in incomes of the unemployed, particularly arising from (2) and (3) above;

(5) mobilize the poor so that they can gradually become aware of their rights and privileges and look after their own interests; hence the need to create built-in systems, within the given socio-economic structure of a given mixed economy, for participation of the poor in decision-making, implementation and an equitable sharing of the benefits of development programmes and projects executed with their labour;

(6) self-reliance, or self-sustaining development at the community level to be achieved mainly through the capitalisation of labour;

(7) provide adequate "working capital" for use of labour as capital, as stated in (6), in development programmes and projects undertaken at the local level;

(8) provide necessary management and technical know-how to the community at the local level for assistance in designing and management of development programmes and projects;

(9) make special allowance for the development of backward regions, areas and communities, by providing additional facilities and concessional terms for (7) and (8);

(10) find resources to provide essential public services, as stipulated in (3), on an expanding and self-paying basis or without undue burdens on the national budget;

(11) a built-in system of monitoring and on-going evaluation of development programmes and projects, based on their actual performance, in terms of both targets or outputs and impact or objectives; and

(12) a system of administration reorganized or geared to the needs of the alternative "development strategy", that is to say, which in turn will also be evaluated in terms of actual performance on the basis of objective criteria, such as quantity, quality, cost per unit of the outputs, and time taken for production; in other words, performance evaluation of administration will be linked to and based on the performance evaluation of development programmes and projects as stated in (11).

At the top of the list of course, is the need for committed leadership with integrity and vision at the apex of government which will be expected to select persons with similar traits as heads of agencies and programmes established to design and implement a poverty-oriented strategy for development "of the people, by the people and for the people"

If we can agree on this list of basic or major ingredients of a people-oriented strategy for development, we may turn next to a scheme or model which, although based on development experience in a least developed Asian country, may have wider applications for predominantly rural economies of the Third World.* The model or scheme is presented here for critical

* I may however clarify at the very outset that this scheme, even though tried in parts (such as in rural electrification projects in Nepal), has to my knowledge not been tested *on a full scale* in any country so far. The scheme or model has been presented by the author in several forums, national, regional and international, and has been well received

review and analysis mainly from the point of view of its viability and applicability under existing politico-administrative and socio-economic conditions in the respective countries as represented here.

III

A Scheme for Self-Sustained Development at the Local-Level

Given the resource and socio-political constraints for using abundantly available surplus labour on a non-exploitative and participatory basis, the purpose of the scheme is to re-examine existing development strategies, institutions and mechanisms, on the basis of the criteria provided by the package of objectives and remould and reorientate them to achieve those objectives. Therefore a quick recapitulation and review of existing practices of financing and executing rural development programmes and schemes may be in order.

Standard Grant-in-Aid Model

A standard system adopted by countries which have instituted community development or rural welfare or village/rural development programmes has the following features and drawbacks :

---

with several valuable criticisms and comments which have been incorporated to the extent possible. The scheme was originally published by the author in a series of articles in Nepali and English during 1970-72. These were later compiled and published in book form under the title "Rural Development, Why and How?" (in Nepali) by Sajha Prakashan, Kathmandu, Nepal in February 1972. A somewhat similar scheme, at least based on similar concepts, is contained in an unpublished note (mimeo) by K N. Raj, "Employment Creation through a Labour and Development Bank; A Preliminary Note", (6 pages) dated 23 January 1972. This scheme and its implementation and results are also described in *Poverty, Unemployment and Development Policy : A Case Study of Selected Issues with Reference to Kerala*, United Nations, New York, 1975.

(1) Grants-in-aid by the central government usually through the ministry of interior or local government or rural development to local communities for execution of development programmes and projects;
(2) Requirement of a matching contribution, usually half of the total project cost, by the local community, mostly in terms of free voluntary labour and locally available materials, such contributions being deemed as equivalent to "people's participation" in the execution of local development programmes;
(3) Central government grants-in-aid are normally channelled through local government bodies or local political representatives, or local administration (sub-district/district level), the latter being often assisted by an advisory body composed of nominated or elected representatives of the local community, their representation being equivalent to people's participation in decision-making in the selection and implementation of local development projects;
(4) Usually a set of criteria are designed for allocation of grants-in-aid to local communities, but in practice these are seldom applied; for seldom can a central government resist the political pressures likely to be generated by denial of such a grant to any community, even as a penalty for failure to show any results for grants provided in earlier years. Therefore the tendency, and very often the practice, is to allocate the very limited grants-in-aid fund more or less equally among all the claimants, without reference to their past performance or any objective criteria that may have been designed for the allocation of the grant. What is often worse, political corruption may lead to arbitrary allocation of the grant to local "bosses" as rewards for political favours done to the leadership at higher levels of the political hierarchy;
(5) The amount of grant available to any single community, therefore, is usually too small to carry out any meaningful development programme and hence, more often than not, most of the projects are left incomplete, leading to dissatisfaction and disillusionment in

the local community and drying up of the flow of local contributions. The flow of voluntary free labour also quickly dries up when the majority of the community sees the scandalous waste of their labour in ill-conceived and poorly executed projects;

(6) Estimates of project costs and estimates of local contributions in labour, etc. are often inflated with a view to seeking larger grants-in-aid; this is yet another reason why most of the grant-aided projects are likely to be left unfinished for lack of adequate resources;

(7) The political character of the grants-in-aid programme, coupled with its being channelled through the political system at the local level, i.e. the unit of local government, leads to a lack of any strict accountability in the expenditure of the funds. This is likely to encourage misuse of funds and corruption at the local level with adverse consequences on people's participation and credibility in government-sponsored programmes, no matter how well-meaning these be, at the local level. The corrupt political bosses at the local level, in turn, blame the central government for the paucity of funds accounting for the non-completion of projects, thereby further accentuating the alienation of the people in the rural areas from the government at the centre;

(8) Even when some projects get finished as planned, the chances are that their lives will be short owing to the lack of maintenance, for usually there is no institutional machinery charged with the responsibility for regular maintenance of those completed projects;

(9) Several projects, under ideal conditions of genuine people's cooperation and honest and enlightened local leadership, do not get finished owing to poor design and execution, because of lack of technical knowhow and management skills;

(10) The total amount of annual grants-in-aid, depending as it does on the vicissitudes of the budgetary allocation, is often unpredictable and tends to remain static even in the face of escalation in project costs.

This list of features and shortcomings of the current practice of the grants-in-aid system is not exhaustive; it is not intended to be so. Its purpose is to illustrate the kind of "slips" or "leakages" that often occur or are likely to occur, in a conventional top-down grant-in-aid programme of rural development. Unlike "mercy", it benefits neither the receiver nor the giver; rather it creates dissatisfaction and disaffection at both ends. It smacks of charity from the high at the top to the lowly down below, or from the centre to the periphery, and it dulls and slowly kills all initiative at the local level for self-reliant development by the people themselves i.e. by their own boot-straps. It even corrupts them into passivity or lethargy and has slowly destroyed in many communities the traditional ways of community self-help, such as in the protection and maintenance of community properties, viz. village forests, roads, irrigation channels, etc. These negative impacts are the very antithesis of the objectives of the strategy of people-oriented rural or community development discussed in the beginning of this paper. Any model of the new strategy therefore must carefully avoid the pitfalls and shortcomings that are likely to arise from only a superficial acceptance of objectives and an uncritical adoption of measures, which violate the principle of equity and may have adverse impacts on the welfare of the target-group.

An Alternative Model

The broad elements of the new scheme are as follows :

(1) Creation of a Local/Rural Development Corporation (LDC/RDC) which will provide medium to long-term (10-20 years) loans on soft terms for the implementation of programmes and projects to be carried out on a self-paying basis by local communities, primarily with the use of voluntary (*not free*) labour. The LDC/RDC is visualised as a domestic or the national version of the International Development Agency (IDA) of the World Bank Group. Its capital fund will be accumulated over the years, if necessary, by the transfer of the amount of "grants-in-aid" appropriated in the annual budget. The net effect of such transfe-

of grant-in-aid funds to the LDC will be that the allocation of these will be handled by the LDC, instead of directly by a government ministry or department. Also the character of the grants-in-aid will change from grants (consumption capital) to loans (investment capital). The capital stock of the LDC may be increased by *ad hoc* grants, and also by channelling funds for food-for-work programmes, and also grants for other welfare programmes.

(2) The LDC, as a banking institution, will be in a position to ask for submission of feasible projects from local communities according to its own requirements. It would also be in a position to select for financing only those projects which it considers as the best or the most likely to succeed. The criteria for selection can be laid on objective bases, and allow for all desirable elements that should be considered in the appraisal of a project for final consideration by the LDC for its loans. Some important criteria for project appraisal, for example, will be the amount of local contributions in the form of labour towards meeting the total cost of a project, or the backwardness of a region or a community. Unlike a regular government ministry or department, the LDC will not be under obligation to allocate funds for every project submitted to it or to every community, irrespective of the initiative shown by it in submitting a project proposal or the soundness of a project proposal. It is true that the LDC as a public corporation cannot be immune from political pressures, but the recognition of its character as a corporate banking institution should enable it to apply its objective criteria in the selection of projects for funding, and also in demanding strict accountability in the use of its funds.

(3) The LDC should be headed by a single executive chairman, who may have an advisory board composed of senior representatives of the government ministries and departments and other development agencies and institutions, such as the Central Bank, Agriculture Development Bank, Industrial Development Bank,

concerned with rural or local development. The LDC should also have competent and experienced staff that can assist local communities in the identification, formulation, appraisal, execution and monitoring and evaluation of projects. The services of the staff should be made available for providing advisory technical assistance to its clientele, namely the local communities as a public service by the LDC. The staff should also guide and encourage local communities to undertake community projects with the LDC's financial assistance. Their crucial role as catalysts and change agents will be discussed later.

(4) The LDC, as indicated under (2), would invite project proposals from all the local communities through their appropriate channels, would appraise them on the basis of objective criteria agreed to by the government, and would be free to choose the ones it considers best for financing through its loans. The criteria can provide for special concessional terms for the poorer regions or communities, such as loans say up to 90 per cent of the project costs, or for longer periods, or at lower interest rates. The policy decisions of the government regarding development strategy in general and local development in particular can be fed into and implemented through the single channel of the LDC, thereby eliminating the need for a slow-moving and too often ineffective machinery for inter-departmental coordination in the implementation of programmes and projects, particularly at the local level.

(5) Projects selected by the LDC for its loans will be discussed with the community or communities concerned in open assemblies with full participation of the people to doubly ensure that the people have had a chance to learn about the project, its scope and limitations, and the implications of its implementation through an LDC loan, some of these conditions, for example, being as follows :

(a) The bank loan will have to be paid on completion of the project in annual instalments until it is fully paid in a specified period;

(b) Although only nominal, there will also be a charge for interest, also to be paid annually after the completion of the project;

(c) The purpose of the loan will be to provide only "working capital" for the execution of the project, i.e. for purchase of essential materials not locally available, such as cement, water pipes, steel rods, etc. and for payment of part of the wages of the voluntary workers for their subsistence;

(d) Voluntary workers willing to work on the project will be entitled to daily wages at the market rate, or if the community is agreeable, preferably higher (20 to 50 per cent) than the market rate. But payment of wages to the workers will be made only partly in cash (not exceeding 40-50 per cent) and the balance in *Labour certificates*, which will be negotiable in payment for use of public services, such as public schools and hospitals, public transport, public utilities such as drinking water, and for acquiring essential farm and other production inputs, such as tools, irrigation water, fertilisers, pesticides. These labour certificates, will be non-transferable, and will be treated as fixed deposits in banks and will earn interest accordingly. In addition, they may also serve as shares in the project wherein the owner of the shares contributed his/her labour, thereby entitling him/her to rights to dividends and representation on the board of management of the project;

(e) Each project thus will be treated as a corporate unit, with a locally constituted board of management viz. a committee composed of the representatives of the shareholders, i.e. the workers who would earn their shares in the project in the form of labour certificates and one or two representatives of the community at large or the head of the local government/ administration unit. This mechanism will therefore entitle the poorest of the poor and the underprivileged (who being unemployed and underemployed will be the ones most likely to work on the local projects) to elect their own representatives to the local project management committees. The homogenous membership of the labour certificate-holding community and the exercise of authority and decision-making powers by them, through their elected representatives on the project management committees, will be instrumental in making them conscious of their rights and privileges (and also obligations) as the majority in a community, and in gradually mobilising them in their own interests;

(f) Since the project should pay for itself, the community will have to agree to the levy of a user charge for its services. The estimated minimum amount of the levy will be fixed on the basis of estimated total costs, etc. The community may however decide to levy higher user charges and repay the LDC loan at an accelerated pace;

(g) The revenues realised from the user charges will be utilised, in the given order of priority, for

   (i) maintenance of the project.
   (ii) repayment of loan instalments and interest, and
   (iii) payment of dividends to shareholders or for partial settlement of labour certificates, the choice being left to the local project management committee.

(h) The LDC will provide one or more technical experts and a manager respectively to execute the project and to manage the affairs on a day-to-day basis including repairs and maintenance of the project and collection of user charges. The services of the project technical experts and the project manager will be made available free of charge by the LDC at least for the initial period of the first five years or so of the LDC's operation in a new community.

(6) A community upon its collective agreement to receive the LDC loan under the conditions as illustrated under (5) will be held collectively responsible for the repayment of the loan, or for repayment of the amount advanced by the LDC in case of abandonment of a project by the community, or non-fulfilment of its obligations such as non-payment of user charges.

(7) Appropriate legislation will also be necessary to enable local project committees to levy and collect user charges on projects completed with community effort, with or without an LDC loan, and also to enable the LDC to realise its loan amount from a community in case of default of payment.

Labour-Certificates

The single most important element and perhaps the most innovative feature of the proposed scheme is the payment of wages to workers employed in rural/local development projects partially in cash for subsistence income, and partially in *labour certificates*. The latter, besides simultaneously giving a labourer an equitable access to public and essential services, future income, shareholders' rights in the project implemented with his labour, would also mop up at least half of the additional purchasing power put in the hands of the newly employed, thereby partially forestalling the rigours of inflation, particularly in the price of the wage goods. The labour certificates would thus also help to raise at least twice the amount of the financial resources available for local level development on a self-reliant or self-sustaining basis.

The concept of the labour certificates may appear to be based on the principle of "work now and get paid later". But the deferment is only for luxurious or non-essential consumption. In case of special needs such as weddings or other urgent social or consumption needs, the labour certificates may also be used as collateral for loans from commercial banks but only against high or penal rates of interest. This avenue should perhaps be left open to the labour-certificate-holders in order not to force them to turn to the moneylenders in case of social or consumption loans. The commercial bank's penalty rates of interest for consumption loans need, therefore, be only slightly lower than the moneylender's going rate.

In order to become acceptable as substitute money, the labour certificates will have to be underwritten or guaranteed by the government or the central bank on behalf of the LDC. The broad outline of the procedures involved may be as follows :

A project manager who will be an official of the LDC will issue a certificate of dues to a worker for labour rendered in a project. This *non-transferable* labour certificate will be deposited in the worker's account with a local bank and a pass book will be issued to the worker by the bank.* The worker will be authorized to withdraw from his labour certificates account coupons in specified amounts or labour certificates in desired denominations for payment of school fees, or payment of hospital bills, or purchase of farm and production inputs at the local cooperative. The government agencies, in turn, will be able to turn those coupons or certificates to their head offices or authorized banks to be credited to their accounts. The non-governmental agencies such as the co-operative societies, or public corporations such as public utilities or agricultural supplies corporation, etc. will be able to credit them to their account with the commercial banks. The coupons will finally end up with the central bank which will make adjustments against the government's account.

Some legitimate questions would be : How and where will the government find the resources to pay for the coupons? How is it different from the goverment or the LDC, acting as its agent, paying in full for the wages of the newly employed in local development projects ? If so, what is the use of the

* The labour certificate mechanism, working through the banking system, will have the additional advantage of expanding the banking system to the non-monetized rural sector in the developing countries.

mechanism of labour certificates and bank coupons? Why introduce them at all?

Subsidies

These are all valid questions. The answer is that the subsidies that most of the governments are providing at present in the form of free education, free medical care, subsidized farm inputs, etc. are burdens on the public exchequer without necessarily benefitting the target group, namely the poor, for which the free or nearly free public services and subsidized essential production inputs are ostensibly meant. However, it is a well-known and established fact in almost all these countries in the region that it is the rich who benefit most from the subsidized public services.

The objectives of both equity and self-reliance incorporated in the new development strategy would also logically imply that the public services must also be provided on a self-sustaining basis.

In other words, non-discriminating or blanket subsidies, like contributions of free labour, should perhaps be abolished once for all, both on grounds of equity as well as for proper accounting of the unit costs of the various public services, and thereby for an appraisal of their efficiency.

Instead, the poor should be enabled to pay for the public services through employment and income, preferably subsidized, i.e. at higher than market rates of wages. This is indeed the essence of the new strategy. That is to say, a target-group-oriented alternative strategy for unified development is to provide the target group with higher or subsidized *incomes* through employment on a *direct basis* rather than expect them to benefit indirectly through use of subsidized public services.

The rationale for higher wages or incomes for the poor, particularly in the form of labour certificates, is based on at least the following four arguments :

(a) the need for equity or for transfer of resources from the rich to the poor in a given community;
(b) to compensate for the inequities in the labour market, in the sense that supplies of labour, except in peak

farming seasons, being larger than demand, market wage rates are normally depressed;

(c) since payment of wages in labour certificates means deferment of consumption and also restriction of consumption to public goods and services, a premium has to be paid in the form of higher than market wage rates to those who would be willing to accept and hold labour certificates;

(d) if as suggested in the paper, fees are to be charged for the public services, there would be need for an element of subsidy in wage rates in order to provide the poor with an easy access to essential public services, which is indeed the primary purpose of the labour certificates.

The purpose of labour certificates, in other words, is two-fold namely,

(1) to subsidize the incomes of the poorest rather than subsidize public services, thereby enabling the present to gain a direct and easy access to the public services, and

(2) to provide a mechanism for the identification or selection of the needy or the poorest of the poor, since it is they who will be willing to work on LDC-financed projects.

The conventional strategy ends up leaving the poor in the cold, both without employment and income and also lack of ability to make use of the subsidized public and essential services. The old strategy, as is well known, benefits the rich and the privileged although, paradoxically enough all in the name of the poor. The new strategy would provide, access to employment, minimum cash income for access to the minimum basic needs, and through the mechanism of the labour certificates, direct access to public services on a discriminating basis that is to say to the poorest of the poor, since it is they who will be willing to work on LDC-financed community projects such as employment (guarantee) scheme on local level projects which thus provides a mechanism for the selection of the beneficiaries or the target group and also ensures, because of the mechanism of labour certificates, that the supply of labour for agricultural operations during the peak farming season, when higher-than-the market wages are paid in cash is not obviously affected. Equity for the poorest, therefore, is achieved directly through employment and incomes rather than through indirect subsidies, which in the context of existing socio-economic structures of the Third World are basically inequitable.

From the point of view of public resources and strict financial management, the labour certificates, turned in for payments, can be adjusted by the government against the savings in expenditures on account of all forms of subsidies for public and essential services. Charges for the public and essential services may be raised if these are found inadequate to meet the cost of the services.

Project Manager

Two other basic and somewhat innovative features of the new strategy also need consideration. These are :

(1) use of the public corporate structure as an instrument of local development, and
(2) coupling of management skills with credit for development.

The first point, namely the use and effectiveness of public corporations as instruments of development is a subject by itself and would require a separate paper. But it is necessary to point out that the basic strength of a public corporation lies in its autonomy and financial accountability within the framework of a given public policy. These two features, therefore, must be preserved in both the LDC and the management committees proposed at the local project level.

It is however the second point that is of a more innovative character, and, based on practical experience, perhaps the single most crucial element upon which the success or failure of the whole new scheme will depend.

That management skills are perhaps more important than credit or other resources has come to be perceived by development planners and workers throughout the world. This perception is based on practical experience and empirical evidence provided by failures of rural credit societies and cooperative societies in several countries of the Third World. In the proposed scheme, the role of the project manager is to serve as the custodian of, or "collateral" for, the funds loaned by the bank (LDC) to the project, and as such he is charged with accountability for those funds. He also serves as the secretary of the project management committee and participates in the deliberations of the committee without any vote. He would however have veto rights against any decisions that may violate the requirements of the loan agreement with the LDC, such as pertaining to payment of loan and interest instalments, rate of user charges, etc. The committee for example may decide to accelerate the payments of loan instalments to the LDC or raise user charges, but it cannot lower them below the rates stipulated in the loan agreement.

The project manager will normally function as the chief executive of a project management committee, and will be charged with the responsibility to implement the decisions of the committee.

The project manager's principal function would be to execute the project, mobilize voluntary labour, make cash disbursements out of the loan funds, keep a record of labour contributions, and issue labour certificates to the voluntary workers and, upon completion of the project, arrange for the collection of user charges, keep account of revenues collected, make regular instalment payments of the principal as well as the interest accrued on the loan, arrange for regular maintenance and timely repair of the project.

The most important task of the project manager however is to generate new projects in full participation with the community, and help them prepare new community project proposals for financing by the LDC, if necessary. The project manager is in effect intended to be a change agent in a local community. His role, in the final analysis, is envisaged to be that of a catalyst in mobilizing local resources for local development.

When one or two projects have been completed successfully by a community, the real significance would lie not so much in the completion of the projects *per se*, but in the discovery or identification of a community that has successfully demonstrated its will and capacity to work collectively and achieve a concrete common asset, for even the most skeptical to see and the villagers to be proud of. The real achievement would thus lie in the uplift of the community's morale, its self-confidence and the strengthening of its cohesiveness. Such a community will be ready to undertake more and large projects. Its success is likely to have powerful demonstration effects on the neighbouring communities. This is the real, although intangible, aim of the new strategy. For, development in the ultimate analysis is in the minds of the people, in the build-up of their self-reliance, and greater awareness of their strength as a community and as a nation. The spark of development thus ignited gathers its own momentum, and like the lamp, it "lights another, nor grows less". Development is thus a process, although a never ending one. It is very necessary to start the process in the right direction, give it a push in the initial stages, and when it succeeds in inspiring the imagination and vision of ordinary men and women, it will not only take off on its own steam but will give them strength to move mountains.

But the start is not easy, particularly in the disillusioned, demoralised, skeptical and ever cynical world of grinding mass poverty in most of the poorer countries of the Third World. Hence it is extremely necessary to ensure that no lip-service-paying or superficial *ad hoc* expedients are substituted for well *thought-through* strategies and schemes for the attainment of the package of integrated objectives. It will be also extremely necessary to carefully analyse and evaluate every measure and every step against the criteria provided by the objectives, in terms of its likely impact on the lives of the target-group and the community at large.

Given the critical role of the project manager as the catalyst or change agent for development by the people at the local level, what is the guarantee that the project manager will perform his formal as well as real functions well? As an employee of the LDC, and hence only on a temporary assignment in a community, he is not tied by any allegiance to the community with which he may be working at a given time. What are the motivations for him to do his best for the community or fulfill all the functions that are expected of him?

Performance Appraisal of Project Managers

In order to motivate its project managers to give their best to their jobs, an objectives-oriented system of performance appraisal will have to be instituted by the top management of the LDC, particularly since the LDC is bound to grow larger as it becomes increasingly successful.

The essential philosophy and elements of the proposed performance appraisal system may be enumerated as follows :

1. The purpose of an appraisal is *relative comparison* and *not absolute measurement.* In other words, the purpose of any appraisal system should not be to rate employees by themselves (absolute appraisal) but grade them *in relation to* other employees of comparable rank or level (relative appraisal or "peer comparison"). This seemingly simple and obvious concept underlying any system of evaluation or appraisal is often missed, particularly in the public sector bureaucracies.
2. Appraisal of an employee should be done in terms of his/her "performance", that is to say, his/her achievement in terms of net output in relation to the goals and targets of the organization to which he/she belongs. The output should be appraised, for purposes of "peer comparison", against suitable *objective* criteria like quantity of production, quality of the product, cost per unit and time taken for production. Often a simple criterion such as ratio of overhead costs to total turnover, for example in the case of a trading company, will be adequate and more meaningful for purposes of the relative evaluation performance than a host of criteria, based on personal attributes of an employee, as appears to be the normal practice in most government bureaucracies.
3. The criteria for relative performance appraisal or peer comparison again need not be perfect but should be adequately designed in terms of an organization's objectives, and what is more important is that it should be *uniformly* applied to all employees whose performances are being compared. (There can,

for example, be no end of debate if it were to be discussed whether there should be 36 inches or more or less in a yard. That need not be the relevant issue in the acceptance of the yard as a standard for linear measurement as long as the yard is universally accepted as the standard and is applied uniformaly, i.e. without variations in its length to a given set of customers.)

4. The norms for performance may be set on the basis of the peer-group's average output in a year or any other relevant basis as suggested in 3. above. All the employees need to be ranked in order of their performance at the end of every appraisal period. The career advancement or promotion of employees should be based on their consistently higher-than-average and relatively better performance than any other in the peer group over, say, a minimum of three to five years. Similarly employees consistently performing below the norm set for the group, such as the group average, may be penalised in terms of their career advancement, including termination of their services if they consistently remain at the bottom of the scale.

   In the case of the project managers of the LDC, their evaluation performance may be appraised in terms of the costs and quality of projects completed, number of new projects initiated, new proposals in the making, mobilization of local mass support, clientele satisfaction, and so on. A strict system of regular performance appraisal will have to be instituted and impartially adhered to along the lines suggested in paragraphs 3 and 4 above.

Some other issues related to the administration of labour certificates are given at the end of this paper.

Conclusion

An attempt has been made in this paper to give in bare outline the essential features of a scheme for an alternative strategy for development in the predominantly rural economies of the Third World.

The focus of the strategy is at the local or village and community level development.

The strategy incorporates the commonly accepted objectives of growth-with-equity-oriented participatory development for employment and basic minimum needs focused on the target group, namely the poor, comprising the small farmers, landless and village artisans.

The strategy is based on the use of investment of unemployed and underemployed abundant labour resources on a self-sustaining basis. This perishable resource unfortunately mostly goes to waste, for lack of both management skills and financial resources to put them to productive use in the labour surplus economies.

An essential and innovative feature of the new strategy is the use of labour certificates for capitalization of labour without its exploitation. The labour certificates will provide to their owners access to public services. The latter, often provided on a limited basis, again because of constraints on the public sector financial resources, need no longer be subsidized and can be provided on a self-paying basis. Hence these services can be expanded to meet the increasing demands of the poor who would now have the means to pay for those services.

The use of labour certificates will remove constraints of financial resources and also reduce the dangers of inflation, particularly in the price of wage goods, as the consequence of a massive induction of purchasing power in the hands of the newly employed. Labour certificates, if consciously and judiciously used, may very well become the means to unleash the vast reservoir of creativity and resources available at the local level, as indeed an implicit assumption underlying this paper is that the "people's sector" comprising the majority of the population ought to be not only the largest but also the most productive sector in an agricultural "mixed-economy".

The labour certificates can also be used as fixed deposits for interest earnings as well as certificates of shares in projects completed with their owners' labour and thereby provide access to positions of decision-making in project management.

The labour certificate is essentially an instrument or a

substitute currency, and although its use will require new institutional and legislative arrangements as indicated in this paper, the latter need not necessarily follow the model given in this paper. Instead of a new LDC, for example, an existing institution like an Agricultural Development Bank may also be entrusted with the task of giving long-term loans at low rates of interest, coupled with the services of technical experts and project managers, for programmes and projects to be executed with voluntary labour in the rural areas. Similarly, existing structures of cooperative societies may be adapted for introduction of labour certificates, although such a decentralized (local) arrangement for introduction of labour certificates is likely to reduce its effectiveness and also impair its general acceptability.

Again, if an LDC/RDC were to be established, its funds need not necessarily come from the annual appropriation for grants-in-aid for local level development. In practice it may even be difficult to make any cuts in such grants-in-aid, no matter how unsatisfactory their use may be, because of political and other pressures from vested interests. It may therefore be necessary to create a capital fund from soft loans or grants from international aid agencies or to transform existing food-for-work and public work programmes into a rural development fund.

The use of labour certificates as certificates of share in the completed village projects if considered not feasible in some countries, at least in the initial stages, need not be introduced in the very beginning, and may be added only later after the labour certificates become generally acceptable to all concerned.

The purpose of these illustrative examples is to indicate that certain institutional arrangements suggested in the model given in this paper may be modified or withheld within the framework of given societies. It however needs to be stressed that any modifications or new measures that may be introduced must be carefully examined in terms of the objectives of development and their impact on the target-group. This is the fundamental principle that should never be lost sight of. It constitutes the essence of a unified or integrated approach to development.

ADDENDUM

Points raised in discussions on the paper in various forums have mostly dwelled on the operational aspects of the labour certificates and related issues, e.g. LDC, subsidies for public services etc. An attempt has been made to classify and the related questions under broad headings and provide answers to them.

LDC

If the LDC were to operate in every village/community would it not mean a very large bureaucracy ? What would happen to people's participation and self reliance ? Would LDC have the technical manpower and administrative capabilities to cater to all requests for assistance in local level projects, particularly in small and remote communities ? Again, will not the creation of the LDC add one more tier to the already existing heavy bureaucratic structures which use up most resources and leave very little for actual development work in the rural areas ?

The fears that the LDC may grow in a monolithic bureaucratic organization are genuine, and this may happen over the years, in case the leadership was not always alive and alert to keep the role of LDC as a catalytic agent rather than a doer of local level development constantly in view. Much will therefore depend on the leadership of the LDC. Therefore, it is stressed that extreme care must be taken in selecting the executive head of an LDC and that he should also be allowed a free hand by the political leadership in a country. There would be a great temptation to the political leadership to fill the position with a political henchman to serve its own political interest. That would be most unfortunate. But that would also be a test of the "political will" of the country's top leadership.

Given the political will at the top and selection of a well-motivated, qualified professional executive head of an LDC in a given country, it is envisaged that the LDC will begin in a small way with a few projects to start with and gradually build up its staff expertise and resources, and therefore its capacity to undertake gradually a larger number of projects. The emphasis however would always have to be on quality rather than the quantity or the number of projects completed. That should perhaps be the main criterion for the evaluation of an LDC's operations.

Again, in a large country where there may be need for more than one LDC, even though one or two LDC's may be established in the beginning, more may be replicated on a regional basis. In a country like India, for example, each State may have one LDC of its own.

Although in the paper it has been indicated that a change agent from an LDC should remain in a community for a relatively long time, particularly after the successful completion of one LDC-aided project in order to promote more community projects, one of the projects of the LDC and its change-agent in a socially-motivated and development-oriented community should be to set up its own institution or agency for local development, such as a cooperative society which may then gradually take over the management of community projects and thereby release the LDC and its "change agent" to work elsewhere. Helping to develop local institutions for community development is only a legitimate goal of the LDC and should be one of the criteria for the performance evaluation of the LDC as well as its community workers or change agents.

Regarding the point that the creation of one LDC may add to the existing heavy bureaucratic overhead costs, we may look at this issue in two ways.

(a) If the recommendation made in the paper, namely abolition of all subsidies to various public services were to be accepted, a sizeable reduction in the existing bureaucracy administering a host of known and unknown subsidies may be expected.

(b) But even if it were politically unacceptable to withdraw all or some of the existing subsidies or to reduce the bureacracy administering those subsidies (even if the latter were to be removed), a policy decision could be made not to extend further the scope of the subsidies, new or existing, or limiting such subsidies to only a few selected services, such as primary education or primary health-care, public transportation etc., and removing from other services such as electricity, fertilizers etc. This is further discussed below.

Q. The administration of labour certificates—issuing them, keeping their accounts etc.—could be a costly affair and may require a large bureaucracy of its own.

A. The labour certificates may be printed and issued by the Central Bank on behalf of the LDC or the LDC may issue them by itself. The administration or account-keeping will be done through the banking system, in the sense that all labour certificates will be credited to the personal accounts of the owners with the local branch of a commercial bank and will be treated as fixed deposits. Banks will issue coupons or certificates for required amounts for payment of school fees or electricity bills for all approved purposes, at the request of their clients. Similarly, the public service agencies such as schools, hospitals, public transport companies etc. will turn all their coupon or labour certificate receipts into their accounts with their banks for encashment. (The commercial banks, as indicated in the paper, in turn will send them to the Central Bank and its branches for credit to their accounts with the Central Bank).

The administration of labour certificates through the banking system would imply that commercial banks would have their branches in rural communities. This assumption may not be valid in all countries, particularly in the least developed countries where banking may not be well developed. Development of rural banking is assumed in the paper as a desirable objective (or a "means" in the "means-objectives" hierarchy discussed in the paper) of rural development. The scheme of labour certificates is expected to initiate the poorest and the majority of the rural population into banking habits, and also to inculcate in them habits of saving through the banking system; all these measures in turn being conducive to expansion of banking to rural areas.

Q. The labour certificates may be sold in the black market and thereby the whole purpose of benefiting the poor by providing access to them to the public services may be defeated, besides leading to corruption and exploitation of the poor.

A. The labour certificates will be issued in the name of the workers and will be non-transferable. Also their operation only through the banking system will ensure that unauthorized persons will not be able to (mis-) use them.

Q. Since the labour certificates can be earned by anyone willing to work on a community project, its benefits (in terms of higher wages and hence subsidized access to public services) may accrue even to the middle or higher income families, and hence may not necessarily be restricted to the poorest of the poor.

A. It is true that labour certificates may be earned by anyone willing to work on community projects. But it would be logical to expect that mostly the poorest of the poor with no alternative avenues of employment will be attracted to the jobs that will offer part of the wages in labour certificates. These job opportunities may be seen as of the last resort. The very nature of the wage payment will thus act as a selection mechanism for the poorest. Now given these conditions if some middle income farmers would still like the work for labour certificates, this need not be lamented or decried, for given the rigid social structures and allied social inhibitions in traditional societies such development may well be welcomed as indicating an attitudinal change in the richer farmers. For the process of development and equitable transformation in a society, in the ultimate analysis, may be seen as a change in the attitudes and minds of people in the society.

Q. Assuming that the principal function of the labour certificates would be to provide access to the poor to essential public services, how would the poor not earning such certificates gain access to those services, particularly so, when it is suggested that the public/essential services be no longer provided free or on subsidized basis and that a cost-based fee be charged for them ?

A. This is a very important issue that has two implications viz.,

(a) for the use and hence attractiveness of labour certificates, in case essential public services were, for various reasons, to be continued to be provided free of charge;

(b) the more delicate issue of possible discrimination

among groups of poor in case fees were to be charged for the essential services.

The issue, however, will have to be reviewed and answered in the given circumstances of each country.

If the country is already providing free essential services, such as primary education and health-care, and is not ready for various reasons to charge a fee for such services, labour certificates may be used to gain access to higher education and other possible public services and utilities such as electricity, irrigation water, transport services etc. This will still imply that there will be discrimination in favour of those that choose to work on community projects rather in the private sector or wholly government financed public sector projects. Such discrimination need not necessarily be regarded as bad or undesirable and may be regarded as an element of premium for foregoing current consumption of wage goods etc. by accepting part of the wages in labour certificates. However, care would have to be taken in not fixing the premium (or the differential between-part-cash-part "labour certificates" wages and only cash wages) too high that there is an unduly large shift of farm labour even during the peak agriculture season cash wages are normally the highest. This is a consideration that an LDC and local communities will have to keep in mind while fixing local wage rates with the labour certificate component.

REFERENCES

1. Such as embodied in "What now? Another Development" (Dag Hammarskjold Foundation) Mahbub ul Haq's *The Poverty Curtain* and the two studies sponsored by UNAPDI, namely.
   Wahidul Haq et al., "Towards a Theory of Rural Development" 1975, and "An Approach to Micro-level Development : Designing and Evaluation of Rural Development Projects," 1977, published in *Development Dialogue* 1977/2.
2. *Rural Asia : A Challenge and Oppotunity*, Asian Agricultural Survey, 1976, p. 15.
3. *The Employment Situation in India, Bangladesh and Sri Lanka. An Overview*, Marga Institute, Colombo, January 1976.

TWO

PARTICIPATORY DEVELOPMENT
The Need for Structural Reform and People's Organisation

A.Z.M. Obaidullah Khan

Development Defined

"DEVELOPMENT is the development of man, as an individual and as a social being, aiming at his liberation and at his fulfilment. Development should be geared to the satisfaction of needs beginning with the eradication of poverty ; indigenous and self-reliant that is relying on the strength of the societies which undertake it ; and in harmony with the environment".[1] The reduction of hunger, poverty, unemployment and inequity together constitute a test of development but the objective of development is to improve the quality of life of the masses of people and unleash their creative energy. Equally important, therefore, with the basic needs of food, employment, health, habitat and education are participation by the majority population in decisions that affect themselves, their family, their community and their role in society at large.

People are the focus of development and of them the poor majority. The bulk of the people live in the countryside and they are overwhelmingly peasants—tenants, share croppers and at the bottom, the landless agricultural labourers. Development, therefore, is not the development of an area arbitrarily called rural nor is it development of agricultural production with a component of improving social services. It is a fundamental process of social, economic and political transformation of a peasant society in which the main actors are the majority members of that society themselves. Development is not just increasing an abstract aggregate measure of national income. It "has to do with the whole shape and texture of human existence : it is not the size of the economy, but the quality of life for each member of society which is our final concern".[2]

Development is not "catching up" with the industrialised west through capital and technology transfer ; it is indigenous to each society and builds primarily on a country's own resources—mainly land and people. Each country thus seeks its own path to social fulfilment and enlarges the capacity of individuals and communities to create and innovate. Development is not for the *elite and the best* ; it is for the many, the rural majority. It is also by the many.

The present crisis in deepening poverty is not merely an unfortunate episode, "but an inevitable consequence of past intentions to exclude the rural majority from development planning and processes, to avoid institutional and structural reforms including agrarian reforms and to concentrate narrowly on increasing production. The present tendency to focus discussion on the so-called "target groups", the rural poor, is to be welcomed in that it emphasises that rural development should be primarily for the poor. It is incomplete in that it fails to emphasise that development be increasingly by the poor".[3] This implies the involvement of the rural poor in the development process and requires their participation in decision-making and implementation of those decisions.

Role of Government and External Agencies

Governments and policy makers at the national level do have a critical role to assist in this process of change through resource-allocation decisions, rural-urban terms of trade; appropriation and reinvestment of surplus in the countryside to give increased economic opportunities to the majority population ; a broad-based educational effort ; an appropriate industrialisation policy that supports the central mode of production i. e. agriculture and offers productive and remunerative employment to the majority members of the community ; and through structural reforms that provide increased access to productive resources by those who have no resource or only marginal access to such resource.

External agencies can assist Governments in this task. But neither should think that they are "doing" development. *Unfortunately*, what we are experiencing today is *development administered to the people.* It is the policymakers, planning bureaucrats and technocrats who either along with external donor agencies or by themselves decide what is good for the rural people. Then they try to impose that package on the rural population. Such rationally designed packages often go beyond the understanding of the rural actors, inhibit their indigenous creativity and thereby make them more confused and dependent.

Self-reliance

Central to the approach for participatory development is the concept of *self-reliance.* Self-reliance at the individual, community and national levels does not call for autarky, but requires that interactions between actors at the three levels take place on the basis of each starting from a position of near-equality, so that the outcome is interdependence rather than dependence of one on the other. "The mobilisation of the energies and resources of the rural poor themselves emerges as the key-factor in increasing their productivity and self-reliance. Such mobilisation requires the formation, adaptation and strengthening of community structures including organisations of the rural poor"[4] and the women. Such organisations depend to a large extent on how power is distributed among different social groups. The ability to mobilise local resources, and voluntary sacrifices for the benefit of the community and the country depend on popular confidence in the belief that benefits as well as costs will be equally shared. Unfortunately, agrarian social structures in the Third World give few grounds for such confidence. The distribution of land is often highly unequal, while the relationships between social groups and the distribution of power are primarily a function of land ownership or the relationship to land in the case of tenants, share-croppers and landless agricultural labourers. Access to water, credit, technology, extension services, education etc., and thus the potential for income growth, are all closely related to the amount of land under one's control. Such stratification creates relationships of dependency between social groups, for instance, landless labourers and poor peasants depend on the rich or middle peasants to give them employment ; small-holders and share-holders and share-croppers depend on the large land-owner to lease them additional lands; all the poor groups depend on the money-lenders, frequently the large landowners themselves, for loans to cover their families' consumption requirements, for production purposes and for special ceremonies. Stratification also influences the amount of family labour available for production, since the women, and sometimes the men, of certain social groups may not work in the fields. Therefore, for any broad-based participatory developmental process, the conflict of interests between different groups in agrarian society must be recognized, understood and taken into account. In such a stratified society, setting up organisations for the whole community may perpetuate the tenant debtor dependency relationship. Yet how can the marginal farmers, share-croppers or landless labourers organise themselves separately ? Without organisation they are powerless to have access to the services and exchanges provided by the government or to express their needs and mobilise their resources. On the other hand, they are dependent on the few and the powerful. In that context, a broad-based participatory development requires reduction in inequality of land ownership. One condition for broad participation is that the majority has access to productive resources be it land, water, credit or education.

Land Reform

Land ownership in many of the Third World countries is synonymous with control of wealth, social prestige and political power. Redistribution of landed property in favour of landless workers, tenants and small/marginal farmers would contribute directly to the alleviation of the most acute forms of deprivation of the rural majority and reduction of their dependent status. As the ILO document on Poverty and Landlessness in Rural Asia points out, "A reduction in equality brought through a redistribution of landed property and supported by measures such as the provision of credit and marketing facilities is likely to raise total production and is certain to raise the incomes of the poor. Rural development cannot be said to have begun without land-reform".* The conclusions drawn from an analysis of the historical experience of selected countries of Asia are that a redistribution of productive assets, notably land, is likely to be an essential component of any development

* "Poverty and Landlessness in Rural Asia" ILO, Geneva, 1977.

strategy that builds on the rural majority and that such reform, in many cases, is likely to result not only in greater equality but also in higher output and greater savings for investment.

The above are borne out by the radical land-reform that has taken place in countries and regimes as ideologically divergent as the Peoples' Republic of China and South Korea. In the case of South Korea such redistribution was imposed from above and eliminated a wide disparity in the countryside between those who controlled assets and inputs, including credit and those who did not. The beneficiaries were then organised into cooperatives for the channellisation of inputs and marketing of output.

The Chinese land reform followed the victory of the party through a nationwide peasant revolution. There was thus an identity of interest between the goal of the national leadership and the demand of the peasant majority. Land-reform was the first step in the redistribution of political and social power in the villages for a thorough institutional reform in future. In carrying out the land-reform the national leadership did not take the simple way of merely relying on administrative decrees and bestowing land on the peasants. It encouraged the peasants at the village level to organise themselves (Peasants' Association, Women's Association) to take the land themselves and redistribute the same. The participation of the peasantry through their own organisation was essential to ensure that a substantial initial redistribution occurred and to set in motion a self-propelling force in which land-reform could become a stimulus to further progress.

Second, the local organisation of the poor peasants threw up a vigorous local leadership cadre that got committed to a new way of life. As Mark Selden notes, "For a generation of youth in North Shenshin, participation in land-revolution was the decisive act in breaking the ties that tied them to a decadent and oppressive order and in defining their involvement and leadership in the new society".[5]

Third, land revolution in China liberated for the first time the Chinese women who got an equal share in the redistributed land, organised themselves into women's associations, joined the peasants' associations and stood up to the old landlords in the village meetings.

Fourth, land-reform raised the consumption level of most of the peasants and at the same time increased rural savings for local and national investment. According to John Gurley "land-reform probably contributed over one third of the total savings in 1952 to the investment programme".

As contrasted to the thorough going redistribution of land and assets, there are examples of legislative land-reform in India and Bangladesh and limited land-reform in Egypt, Philippines, Thailand and Pakistan.

The legislative land-reform of India and Bangladesh attempted to by-pass the local power structure, was unclear in definitions, uncertain in implementation and piecemeal in efforts. The implementation was essentially bureaucratic without an alliance between the central and state administration with the peasantry. As Daniel Thorner states, "The small minority of oligarchs have had wit and resource enough to get around these laws, in which in any event the loopholes were so large as to give them ample manoeuvring ground. In Indian conditions, if you do not totally reject the principle of non-working cultivators, you cannot prevent the village-oligarchs working as landlords. Lands and fruits thereof are to belong to those who plough, harrow, sow, weed and harvest. This will result in a major redistribution of income, to the advantage of those who work in the fields. Control of a very considerable amount of land will pass from rentreceivers and other soft-handed folks to tenants, share-croppers and labourers. The agrarian transformation must come first. Only then can there be a proper footing for the long needed rationalisation of agricultural production. Until proprietary rights are ended, cooperatives will not stand a chance of working out successfully.

"The programme of abolishing proprietary rights is clearly and easily understandable in village-terms. Every peasant in the village knows who does what kind of work, under what terms, in whose fields".

"Of all demands it is one made in most parts of rural India by the working peasants themselves. The heart of power, prestige and standing in the village lies in land. Put land in the hands of those who are working it and you crack the existing concentration of power".[6]

The analysis of Akira Takahashi of restrictive land reforms in Pakistan (1959 and 1972), Philippines (1972) and Thailand (1975)[7] also concludes that a programme which does not distinguish between the beneficiaries and in fact favours the better-off sections of the rural population at the cost of the marginal farmers, share croppers and landless labourers may be an important contributing factor to rural poverty along with population-growth and lack of alternative employment opportunity. Further in the case of the Philippines the absence of a close alliance between the administrative machinery for implementation and the tenants and the peasantry, has resulted in "physical acts of dispossession, like bulldozing of farms, demolition and/or burning of houses, man-handling, intimidation etc." by the local power-holders, the landlords.[8]

The agrarian Reform Law of 1951, 1961 and 1969 in Egypt succeeded in redistributing 13 per cent of the cultivated area to about 9 per cent of the rural population. It eliminated the very large land ownerships (1.5 per cent of land-owners) and consolidated the position of the rich and middle peasants which represents the economic and political power in rural Egypt, and slightly improved the lot of small peasants for a relatively short time-span. The initial inequality of land-ownership was not changed in any fundamental way. The landless peasants remained outside the reform and in 1972 they represented about half the rural population. There was a significant fall in the absolute number of landless families between 1950 and 1961 but the 60s and the 70s saw a reversal of the trend partly due to the population pressure on land and partly due to the apparent freeze on agrarian reform. The marginalisation of small peasants and continued concentration of land in the hands of medium and large land-owners, together with a high rate of population growth have already and will continue to cancel out the effects of any marginal land redistribution programme.[9]

The dramatic increases in rural poverty in some of the countries mentioned above (44% of the rural families in Egypt were living below the poverty line in 1974-75 as opposed to 27% in mid-sixties) can partly be explained in terms of the failure of agrarian reform policies to alter fundamentally the unequal distribution of land, income and power in the countryside.[10] The marginal redistribution could lead to some temporary improvement which is soon overcome by the dynamics of the situation that reproduce inequality.

Similar reproduction of inequality can also be seen in land-settlement operations in land-rich countries like Malaysia. The pattern of land ownership remains unchanged, the confrontation with the vested interest is avoided and an attempt is made to create a prosperous peasantry on the margin of the new land. As the ILO document on Landlessness and Poverty in Rural Asia points out, a severe limitation of policy was its lack of focus on specific groups. The situation of the marginal paddy-farmers outside the "rice-bowl" areas was overlooked as was the predicament of the rubber small-holders. The settlers (to the new land) were selected on the correct principles of equity, but the enormous imbalance between the number eligible for settlement and the number of settlers who could be accommodated on these schemes was overlooked. Further the original inequalities in income distribution on the old land remained unchanged. The various land-reform measures taken in the countries of Asia except in the cases of thorough going reforms in Japan, China, Taiwan and Korea have generally strengthened the rights of a minority of middle-level tenants and owner-operators, together with increased productivity in the short run as a result of the delivery of technological inputs to them. The consequence, often, is the emergence of a new class of share-croppers, sub-tenants and seasonal workers nullifying the little which has been achieved under the reform programme.

The growing problem of landless labourers adds a new dimension to agrarian reform, which has almost entirely been neglected, except in terms of enacting minimum wage-laws which cannot be enforced. Irrespective of the form of reform, given the prevailing man-land ratio in many Asian countries, it is essential that the reform programmes entail fuller utilisation of the growing mass of landless labourers whether on-farm or off-farm. It is often argued that in countries with high man-land ratios, redistribution of land directly to those who have marginal or no access to land may leave out yet a large number from such access as there is not enough to go round. Even the demand for the labour of those who are left out may decline because of fuller utilisation of the family-labour by beneficiary households. But, if such reform is taken as a first step for

future co-operative farming, whether in the private ownership context as in Japan, Taiwan, South Korea or in the communal context as in China and the Kibbutz of Israel, the problem of miniscule fragmentation can be overcome. Secondly, such cooperatives or associations linked with local organisations of appropriate scale could comprise a package of activities aimed at the fuller utilisation of the available labour-force and infra-structure (roads, irrigation-channels, storage godowns, extension and credit-services, community education aad health-services etc.). Again structural solutions in such countries could combine with diversification of rural activ ties including communal forestry, fishery, livestock, a public works programme for strengthening the rural infrastructure, side line occupations like handicrafts and local-resource-based industries. Such an approach would be in marked contrast to the present land-reform laws which have ignored even the existence of the problem of the landless workers.

Further, as C.L.G. Bell and John H. Dulay point out even if the amount of land that can be redistributed is limited, making plots available for house and kitchen garden go a considerable way towards reducing the dependent status of landless labourers.[11] Similarly, the Food Crisis Workshop held in Manila (Feb. 7-9-1977) by the Ramon Magsaysay Award Foundation states, "The greatest insecurity in most Asian villages is ownership of house-lots where rural families live. For example, in the Philippines less than one half of all families own the lots on which their houses stand and, therefore, are discouraged from making even simple improvements like installing a water-sealed toilet. Ownership of a house-lot of 600 to 1000 square meters gives a family security of belonging in the community. It also gives the family title to a piece of real estate with improvements that can be mortgaged, should a loan be needed. For Asian governments the opportunity to carry through a land-reform assuring every family of the ownership of house-lot affords the most dramatic wide-spread and cheapest means to rapidly achieving rural well-being, and greater prosperity". Further, plots large enough to support a kitchen-garden, a cow or a buffalo could be of considerable nutritional value to the family. In the People's Republic of China, the private plots provide the vegetables and protein needs of the rural families (i.e. pigs, poultry, fish) and complement the basic food-stuff provided by the collective.[12]

Within the above spectrum of countries and policy-options, there is, however, a consensus in development thinking today that an equitable and productive land-system is a necessary condition for the initiation of participatory development with benefits widely distributed.

Again the land-reform process itself could generate peasant enthusiasm and peasant organisation. Greater equality of incomes has often been associated with extensive participating rural organisations not merely as a receiving structure for inputs from the top, but also as actively interacting with higher levels of decision-making for mobilizing local resources as well as claiming outside assistance on a consultative basis. The examples in the Asian context are China, Israel, Japan, Korea, Taiwan and Yugoslavia each with its relatively organized local communities and with varying degrees of autonomy in local decision-making vis-a-vis the higher levels of governmental apparatus. What most distinguishes these cases is effective and serious land-reform, not just breaking up some big estates but systematically getting land to small and landless farmers within some framework of rural local organisation.[13]

The importance of effective local organisation for improvement of agricultural productivity, diffusion and adaptation of technology, diversification of agriculture, increasing employment opportunities, improved education, nutrition and health care are borne out by the experience of the countries mentioned above.

One common denominator which stands out among countries with successful rural development experience is the multilevel linkage of such local community organisations and the balance between solidarity and the requirements of scale.

At the base is an organisation small enough for shared identity and common interest. In a production team in China everyone knows who contributes well and who poorly, with the former rewarded by local esteem and the latter censured by local pressure. The production teams in turn are woven into the production brigade which can mobilize the labour force for infrastructural work (i.e. irrigation canals, maintenance and repair, electrification, roads, storage etc.). The commune provides administrative, health, educational and industrial infrastructure, and thus builds up an agricultural-industrial-cultural local economy. The Japanese *muro* (hamlet) with *buraku* (administrative village), the Taiwanese township with its inter-village farmers' associations, and irrigation associations, the Yugoslav commune with constituent neighbourhood associations correspond to such multi-level local organisation. What is important is the interdependence between such organisations and the administrative apparatus of the states. Ideally, if the rural development objectives of the centre are consonant with the interests of the locality, the ability of the centre to achieve its aims will depend in large part upon the power of the localities. Similarly, the capability of local communities to accelerate their development is enhanced by central government capabilities in the area. Thus, as long as the developmental goals are compatible, central and local governance are inter-dependent and mutually supportive. Within the above framework, the countries which are doing a better job than most countries in improving the conditions and opportunities of the disadvantaged classes are those with substantial amount of decentralisation of responsibility for development to the local community.

No community, of course, can challenge with impunity the broad policy decisions of the top. But within such boundaries, much of the problem-solving innovation is expected to occur in the local community. Should crop-rotations be changed? How can farm productivity be increased? What irrigation technology should be used? Development goes forward in the strong light of community surveillance with a very high percentage of participation and cooperation and a very strong pressure for everyone involved to live up to each others' social expectation. This is not to imply that local engines can run without fuel. Without seeds and fertilizers the farms cannot do what they are expected to do ; without contraceptive materials, the birth rate will be less well controlled ; without a committed and responsive local leadership, local initiatives may go in any number of directions. But in the absence of adequate material incentives in resource-poor countries, it is these social relationships and the psychic income derived from them that provide the incentive for the considerable efforts asked of the people to develop themselves on a self-reliant basis.

Local Leadership

In such a context, the local leadership is the most important intermediary between central guidance and local initiatives. In China 20 million such local cadres carry communication both up and down, to and from top leadership and horizontally in the village, neighbourhood and the commune. They are supervised from the top for understanding the "big picture" but are surveilled from below by the community organisation and by actually working with the peasants and workers in the farm or in the workshop. They are responsible for not special problems but for the whole of development in the locality. There is not (as in most of the developing countries) a vertically linked specialist who comes from the centre or the region to help the farmers, a family planning worker who comes to persuade the families, a health-expert who comes to teach people to dig wells or boil water. The technical specialists come and work with the local peasants and workers to build up local capability. But it is the local leadership which manages the central task of building up a local agricultural-industrial-cultural economy where everybody has a responsibility for it and takes part in it. Similarly, bureaucrats and technicians from higher organs of government are sent down to the commune to work and live with the peasants, thereby relating their development assignments to local perceptions. Development is not for planners, or bureaucrats or leaders only, it is for everybody.

In many developing countries, however, local institutions have been initiated by various development agencies in order to carry out functionally specific programmes. Such organisations are regarded as extensions of the government bureaucracy and are required for delivery of outside resources rather than for mobilisation of local resources and self-management. The extensive supervised co-operatives in Egypt with their pyramidal structure corresponding to the country's administrative hierarchy, has at the base the village-cooperative, the elected council of which is supervised by an agricultural engineer appointed by the Ministry of Agriculture. The same is true of the state-induced co-operatives in the Philippines, which have often been taken over by the more prosperous. "Common interests" so often invoked for whole villages or clusters of villages, in practice become a set of policies mainly reflecting the interests of an activist minority. Such large organisations imposed from above for better management are necessarily run not by face-to-face relationships but by representative committees, and the representative leadership will lmost certainly reflect the dominant social structure and the interests of the dominant group. Even in Bangladesh with the majority of the peasants having a holding of 2 acres and less "combination of the rich, middle, small and marginal farmers together resulted in co-operatives being dominated by vested interests".[14] In Egypt one of the main objectives of the supervised co-operatives after the 1952 land reform was to influence agricultural productivity, through the supply of improved inputs and above all credit. By the 1960s they became the sole supplier of inputs. As the ILO document on "Agrarian Reform and Rural Development[15,16] Egypt, 1952-1975" notes :

> Medium and large landowners found in the credit facilities advanced by the co-operatives a cheap source of finance not only for their agriculture operations, but to enhance their wealth by buying more land, machinery, livestock and sometimes to finance some commercial business. When the Government tried to check this trend by imposing a rate of interest of 4 per cent on loans advanced to holders of ten feedans and more, many large land-owners reacted by dividing their holdings into plots of less than ten feedans to benefit from the exemption from interest.

Unequal distribution of credit was not the only means by which large landowners appropriated the greater part of co-operative finance. A common practice has been to raise substantial amounts of capital through continuous accumulation of co-operative debts. In fact, there were cases where some large landowners assumed the role of the private money-lender, using the credit they obtained from the co-operatives.

Reference must finally be made to the fact that many privileges were enjoyed exclusively by rich farmers by virtue of certain provisions of the agrarian reform laws. For instance, in the system of livestock insurance, "only the owner of at least three head of cattle is formally eligible to insure his livestock, and subsequently to obtain a ration of 150 kg of fodder at the subsidised price fixed by the state. The poor peasant who cannot insure his livestock is deprived of such privileges ; in most cases he has to buy fodder on the black market. Moreover, only owners of more than 15 feedans were eligible to buy selected seeds at subsidised prices. It is not, therefore, difficult to understand the often-expressed view among small and poorer peasants that co-operatives have done very little to relieve their monumental burden of indebtedness."[17]

Similarly traditional local government institutions in many countries, a legacy from the past, do not represent the majority of the rural people, the illiterate and the semi-literate peasants, the landless, the women and even the youth in a hierarchical family structure. Local power-holders can and do manipulate local voting too easily. In the presence of an agrarian oligarchy, giving small men greater influence by "One Man one Vote" does not in fact produce the result. Rural elites, sometimes intimately linked with the national ruling class, are not likely to give up their traditional privileges. They do distort projects/programmes designed to benefit the small peasants or tenant-farmers.

Organisations of the Poor

Some reformist governments at the central level are aware of such distortions and recognise the stratifications in rural society and the importance of "Segregated autonomous politico-economic organisations of the various sections of the poor".[18] Acceptance of such interest-based mass-organisations involves acceptance of a high measure of local political turbulence in as much as they might be regarded as a potential threat to the establishment. Therefore, most major government initiatives may not be expected in this regard in some countries. However, both governmental and non-governmental private sector initiative in promoting such interest-based mass-organisations are in evidence in many countries.

The poor can be organised economically (as in the case of the Anand milk producers' cooperative in India) and thereby develop a counter-weight that offsets their lingering impotence in local government or by formation of restrictive organisations based on interest groups, small-farmers' cooperatives, landless workers' union, women's associations, rural youth brigades, or through organisation of the politically disadvantaged for direct political action as in the case of Tanjore and Kerala in India. In Mauritius, a public works programme in the plantation sector resulted in the unionization of the labour-force.

In recent years efforts at generating organisations of the rural poor have been increasing in Asia. A number of efforts have sprung up in South Asia. The Janavasas (People's Settlements) in Sri Lanka where groups of landless youth have been settled in experimental collective farms; the Bhoomi Sena (Land Army) in India, an indigenous organisation of bonded labour in Maharashtra for asserting their legal rights on land, against bondage and for minimum wage; the TVD (Total Village Development Programme) under the Comilla Academy in Bangladesh in which youth clubs have been organised in the villages for acting as catalysts to organise the landless poor; and the Peasant Asiayone in Burma, a nationwide peasant organisation intended to promote collective participation of the peasantry in Burmese socialist development to mention only a few.

Many more such efforts exist in various parts of the Third World. Together, they are a significant expression of concrete urges of those who seek participatory solutions to the challenges of rural poverty and exploitation. A systematic study of such experience would provide a better knowledge base for rational participatory organisation.

The form of organisation of the rural poor, and the scope and method of its activities, will vary according to specific socio-economic contexts. The production and social relations governing the conditions of work and life of the rural poor are diverse and complex, and no unique or simple type of organisation may be prescribed out of such context. In those parts of the world where the distribution of land ownership is sharply polarised, the landless or poor peasant is often not a simple wage-worker or a single land owner; he may be a sharecropper, casual labourer to many employers including the government, "guest-labourer" to rich households where he eats as well as earns a cash wage, or bonded labourer often with his family to the rich landlord-money-lender. The small owner-peasant in many countries works for others to supplement his income, often works entirely for others and leases out his own land for lack of working capital, and is generally in the grip of the money lender to whom he tends to lose his land and other assets. Even the middle level owner-peasant often loses his land to the money-lender if he lacks staying power to live through unforeseen hazards which often arise.

Thus, questions like exactly which sections of the rural poor will find it in their mutual interest to form organisations for collective action, which other sections they may consider as their close allies to whom they may want even to extend membership of their organisation, and what will be their priorities for collective action, are to be deliberated by the rural poor themselves and understood only against the specific context that prevails in the situation in question. In more egalitarian rural communities, the organisation may embrace the entire community, and will obviously have its own priorities for collective action which may be very different from the pressing concerns of the deprived and exploited in sharply differentiated rural societies.

The task of promoting participatory organisations of the rural poor should not be seen as that of initiating any predetermined form of organisation. It involves mobilizing the rural poor for extent-specific action generated by their own deliberations, which will in turn generate organisational patterns appropriate to the context and action decided upon. Secondly, the organisation will grow and adapt itself to new situations, as the dialectics resulting from organised action itself unfolds and social relations in production change.[19] At the same time, it is important to recognize that these special-interest groups of the poor cannot be effective if they organise simply at the level of the village or peasant community. The reason is that, the constant dependence of the marginal peasantry and landless labourers on the locally affluent and landowning groups, makes them very vulnerable to retaliation, and inhibits the formation of groups organised in terms of economic interest. Besides, the affluent landowning powerholders of agrarian society are either already organized in farmers' associations or similar nationwide groups that constitute powerful lobbies at the state political level, or can easily do so when threatened by indications of the poor's attempts to organise for their rights. For the reason, poor people's organisations must transcend village boundaries. For unless the disadvantaged can increase their political and economic strength through numbers, by crossing village boundaries and joining hands with those in a similar situation in other villages, the contest between the rich and the poor will be a highly unequal one and there will be no doubt as to its outcome.

The fragmented bureaucratic autonomy and compartmentalisation of the agencies often lead to multiple agencies doing the same work adopting different criteria and procedures and consequent waste of resources and, at times, under-utilisation of institutional facilities. The peasants and the rural people tend to get confused and become passive recipients or rejectors. A common example is the Ministry of Agriculture with its functionally specific view-point of increasing production and the newly created rural development or small farmers' agency for delivery of inputs to small farmers. The extension-service and the input delivery system of the Ministry of Agriculture charged with a targeted production goal reach out to the farmers with resources and skill. They also have access to the ministry's field-agents. Thousands upon thousands of small farmers are by passed, in favour of selective contacts, for reaching the production-target. The new agency gets only a fractional residue of the credit or the inputs. Further, the receiving structure created by the agency at the local levels often lends itself to distortion through the manipulation of the local power holders. Another common example is setting up a nutrition cell in the Ministry of Health for school children by the Social Welfare Department which leaves out the majority of the poor non-school-going rural children suffering from malnutrition Yet it is important that the agricultural production process is integrated with the health and nutritional needs of the population and their actual and potential productivity. Further, in trying to reach out to the rural communities there has been a proliferation of bureaucracy clearly beyond the resource capability of some countries. There is a shift in some countries, for example, towards providing health-services through health centres at selected centres. People are now trained in the village to meet the needs of the villagers and an elaborate 3-year training programme has been set up with aid-support in Bangladesh for such paramedics.

The trainees are drawn from the urban centres as 12 years of formal education alone can make one eligible for such training. The medical assistants are vertically linked with the Health Ministry and are not responsible and responsive to the community they are supposed to work for and with. Similarly. a separate cadre of ten thousand family-planning assistants has been recruited. Such fragmentation of health care and population planning is counter-productive and will be clearly beyond the resource-availability for current expenditure by the central Government once the aid-flow stops. More important, a new class of bare-foot bureaucrats is being created divorced from local community, local production and manual work.

In this context many developing countries are exploring different approaches to decentralisation and coordination of developmental activities at region/district levels. What has happened generally is de-concentration of administration from the centre to the regions, the district and sub-district without the turbulence of local politics.

The Bureaucratic Problem

The bureaucratic agents for horizontally coordinating the programmes/projects at the region/district levels are generally responsible to the Ministry of Interior and local government or to the Ministry of General Administration at the Centre. Their basic role continues to be tax/revenue collection and maintenance of law and order. A new role has been added, that of co-ordinating the discrete activities planned and implemented by development ministries on a sectoral/departmental basis. The meshing of tax-collecting and administration roles of the district administrators with the new development-initiator and co-ordinator role results in a bureaucratic style of work which proceeds by orders and constraints rather than by going down to the villages to understand the local situation and working with the local community and departmental agents to obtain positive support for any project/programme. The elite administrator at the district level and the technical experts are generally alienated from the peasantry, often have a contempt for them while professionals such as engineers and doctors do not enjoy professional discussions with the semi-literate or the illiterate peasants. This is very well illustrated in Tschannerl's study of the rural water-supply programmes in Tanzania. As he notes, "the staff of the water ministry tends to have a technocratic and bureaucratic attitude putting primary emphasis on the establishment of technically sound structures, rather than close cooperation with the peasants. Projects are designed, constructed and maintained by the engineers of the water department. Self-help where it has taken place in the past, was nearly always limited to a contribution of free labour under the supervision of the water-department staff"[20]. The involvement of the people in the eyes of the experts will be politics unnecessarily interfering with development work. In such systems, development problems are seen as problems of "administering" things rather than involving the people. Decision-making is typically a movement of files and orders through the hierarchy of officials. The bureaucratic method of decision-making finds its complement in the technocratic method of implementation. The programmes are seen in terms of "correct expert advice", efficient organisation and planning of time, use of qualified manpower, people being just another statistic in the Plan who should implement the instructions. In fact one could make a strong argument that implicit in the process of development is the accountability of the bureaucrats to the people. The bureaucratic style of administering development projects/programmes is partly responsible for the continued remoteness of the rural poor from policy-decisions and consequent inertia, and passive dependence.

Second the absence of any well-defined rural or community development technologies around which professional capacity or resources can be organised casts considerable doubt on bureaucratically implemented integrated rural development efforts, relying on a package of inputs and technologies and strengthening the delivery system. Moreover, given the structural inequalities, it is difficult to design a rural development programme activities package that can command a broad basis of community support.

Third, the efficient delivery of bureaucratic services to the rural community must depend on effective organisation at the rural community level for meaningful interaction with the delivery agencies in the establishment of priorities. Moreover, they must be able to mobilize sufficient political resources to be able to provide incentives for efficient bureaucratic performance. Yet the lack of access by the poor majority in the rural communities to the instruments of central or local governments that would enable them to provide either political or economic rewards for effective performance—by the extension, health or education bureaucracies—reduces the incentive for efficient bureaucratic performance.

Decentralisation of Planning

Effective implementation of rural development depends on the institutional capacity at the community and the governmental levels to mobilize the limited economic and political resources available to the disadvantaged in rural communities. It is possible to organise efforts which expand the resources available to the local majority and contribute to the efficiency of production and consumption activities in rural areas. It is possible to organise schooling in such a way that it contributes to efficiency of the rural people both as producers and consumers. It is possible to train poorly educated village-nominees in health and extension skills that would enable them to deliver, disseminate and adapt production technologies and services to the communities where rural people live and these services and activities can be simultaneously available without being necessarily administratively integrated.

The important question is not that of methodology, but of the fundamental development objective : for whom and more importantly by whom. The village teacher, the primary health-assistant or the agriculture extension agents may not for instance, each be looked upon as the exclusive constituent of a particular department in government or of a particular sectoral agency. Each of these workers, whose work keeps him in almost daily touch with the rural families, has the potential to improve the quality of life of the rural poor, perceptibly even if modestly, as long as they are responsible to the people's organisations rather than to a vertical hierarchical structure.

If we take an approach to development which wants to incorporate more and more people for a better life, the key-resources to be mobilised are the manpower and the existing local resources. The assumption is that their participation in day-to-day decision-making in all phases of local economic and social activities will provide them with an awareness to channel their energies most effectively. Experiences of countries as varied as the People's Republic of China and Sri Lanka, North Korea and South Korea, Japan and Taiwan bear this out. In all these countries, broad-based popular participation has played the key-role in transformation of the agrarian society.

The first step has been some redistribution of current assets, and in most cases, this is land. It could be done by imposition or by involving peasants in local organisation to implement the redistribution themselves. Without some such distribution, whatever we may try will only be a holding operation.

A second aspect is the presence of very strong local organisation, which can deal with the national government on an equal footing. The concept behind this is self-reliance, where there is interaction between the different levels of government on a cooperative and not on a compulsive basis.

Third, a delicate balance between supervision from the top and participation from below has been obtained through the use of intermediaries between government and the people drawn from the community group itself and remaining accountable to the community organisation.

Fundamental structural reform like redistribution of current assets is clearly a political decision to be taken by national governments. There are political, social and administrative constraints to such radical policy choice. Similarly interest-based community organisations in a stratified society have to be encouraged and nurtured by national governments in the face of possible opposition from the local power-structure. What then can be done to promote participatory development even in the often difficult situations which most countries find themselves in ?

The Choice of Technology

The way in which technologies are chosen and applied to the rural area for instance is of crucial importance in the context of popular participation. The basic principle should be one of supporting and building upon the existing technological capability of the local population as well as on the availability of local resources, in order to establish a self-generating process of indigenous technological creativity. The mystique of modern technology as something that is brought into the rural community from the outside, that can be understood and controlled only by a new technocratic class of bureaucrats and research scientists, not only creates new dependencies but also ignores the enormous potential of peasant innovations and resourcefulness.

The constraints of peasant farming and local conditions of labour supply should be the determining factors for the choice of rural technologies, which are therefore likely to be area-specific in their appropriate use. Village-technologies, as an intermediate link between hand-power and energy-and-capital-intensive machine technology, offer considerable scope for future research and development. In the People's Republic of China, research and design teams have been sent down to the communes and brigades to work with the peasants and to consult them about locally applicable technology In the process, veteran workers from local handicrafts units are trained in maintenance and repair; standard blue-prints of machines and machine-parts are widely disseminated and the drive for broad participation encourages improvisations and innovations. Rural industries in China are utilised to supply inputs for labour-intensive development schemes and to produce machines (i.e. hand-tractors, power-threshers, winnowers, etc.) to reduce labour demand at peak periods, thus levelling out the demand for labour throughout the year. In this model of industrial growth, much of the investment in local industries come from incremental production and savings of the localities. These are not isolated in the countryside as there is a dispersal of intermediate-size industries in small cities and country townships to provide a suitable environment and necessary educational fall-out for fostering the growth of largely locally self-reliant small industries.

In the case of the new irrigation-seed-fertilizer technology the process also has to be looked at from the point of view of its accessibility by the majority of rural people. Current high-yielding varieties demand good water control, fertilizers and pesticides. A technology consisting of varieties which thrive under natural rain-fed conditions, which could gain a major proportion of nutrients from organic manures and companion legumes, and which has disease and pest-resistance bred into them would respond better to the technological requirements of the majority in many countries where there is small chance of easy access to high-cost inputs. Such research efforts are under way in the international research centres and in several Third World national centres, and need active encouragement both through funding and training efforts. Distortions in the pattern of benefits resulting from the need to apply high cost inputs could be reduced through strict monitoring of the credit and input-delivery system. However, the move away from simpler, more self reliant technology to that producing a dramatic output increase may entail a cost in terms of majority participation and involve a deterioration in social conditions.

Technology must, however, be seen as more than a means of producing increased amounts of food and consumer goods. The other important human needs of education, health care, habitat and living arrangements (fuel, lighting, water, etc.) should all form a legitimate area for relevant technology. Innovative institutional arrangements will be more important in this area than the actual technology hardware and most organisational components should be provided through training of village/community nominees.

In this context, education and training which enable the majority population to raise their awareness of their position in society and in the economy, to help them organise themselves and increase their political participation is of fundamental importance. This may mean change in the structure and content of education. Yet, support of basic education and on the job training, making education more relevant to local needs, decentralising school administration and putting the schools for basic education under the control of local communities could be the first step which many governments are willing to take. The school must be of the community, for the community and, to the extent possible, by the community. Children should be systematically taught the range of everyday skills used by the community, to learn the crafts from craftsmen, the skills of farming from the innovative farmers and the basic health technologies from local health workers. Each school should strive to be self-supporting as a farm and craft centre advised by the full range of government agents. Basic language, science and mathematics can be well taught in this "real-life" environment and some of these skills will be communicated to unschooled adults within the children's family.

Education and training outside the formal system is best carried out by people nominated by the community for training. Nominees might be given a sectoral orientation, acting as a contact point with field agents in health, cooperatives, veterinary services, public works or agriculture, of a more general training stressing functional literacy and planning skills.

Their success should be based upon the use of local group meetings which are aimed at raising consciousness with a view to analysing the problems and resources of the community (or group) and to planning and implementing self-help projects. This approach is based on an assessment of resource and power inequalities and encourages the community to use available government services, where these are necessary, to supplement self-reliant activities. The approach stresses the collection of information on the physical and socio-economic environment, and group planning and problem solving followed by the organisation and implementation of development projects. It educates rural people to know what services are available from the state, for whom, and how to convey their ideas and problems to the appropriate government agency. Under such circumstances, the community may be fully able to play an active role in testing and research as well as becoming involved in evaluation work on government programmes.

Most government programmes and donor assisted projects have focussed upon training of policy makers, administrators and field agents to the detriment of the training of rural inhabitants. This balance could hopefully be rectified, but the adjustment must be accompanied by qualitative changes in traditional training methods and institutions. Perhaps the most important improvement which should be introduced in the whole range of training curricula is to have a common "core" dealing with the realities of agrarian society, focussing on the unequal distribution of wealth and power, the problems of poorer groups and the experience of most programmes in terms of the unequal access to resources and benefits. Another vital component of the core would be a study of traditional training systems—ranging from child care to apprenticeship of various sorts. The cultural and socio-economic insights obtained in this way should be reinforced by holding analytical sessions on chosen development problems with mixed classes in terms of rank and function.

The Role of Women

As many women as possible should also be included as all these innovations are designed to break down traditional, social, sex and hierarchical barriers in order to gain different insights and to increase the perceptions of the problems of other groups.

It is perhaps a truism to say that women, especially low-income rural women, are probably the most under-rated and under-utilised development resource of the "resource poor" Third World countries. Past programmes and policies for women have concentrated on their roles as reproducers to the exclusion of their roles as producers. Such programmes/projects have involved mainly welfare-services (health, family planning) and instructions in nutrition, and home economics. Even in rare projects where productive skills have been a focus, it is the welfare concern and not development concern that stands out. Thus craft projects which have been developed without market research produce unmarketable items and skills which are sold at subsidized rates, if at all. These programmes, therefore tend to promote the very dependency that they are meant to alleviate.

When concern about women is centered in welfare and special women's programmes, women are effectively excluded from the broader range of programmes and policies, i.e. education, employment, agricultural extension, upgrading of skills for income generation, programmes of credit for cooperative development.

Recently, there has been a recognition by national governments and international agencies of women's economic and productive roles in a community-based participatory development strategy. Such programmes, however, should be based on a clear knowledge of the circumstances of women's lives in various societies, and not on idealised status concepts. In formulating such programmes and projects, their likely consequences on all aspects of women's lives, impact on productive and reproductive roles, the burden and drudgery of their work, their social and economic status, their health and well being, their families, and the community they live in, have to be analysed. Projects for instance, could draw on and seek to strengthen the existing production and other functions and make them more remunerative and pleasant. Women themselves should participate in all aspects of such policy and programme development and implementation.

In many developing countries (Bangladesh, India, Indonesia) landlessness is denying the means of production increases and participation in the economy to growing numbers of people. In that context both on-farm and off-farm technologies suited to their needs in terms of employment income and a decent life must warrant a much higher priority than they currently receive. Massive public works is currently the preferred method of mobilising the labour-force for development of land through capital constructions. It is the socially cheapest way, but with unequal distribution of land-ownership it can result in doubling the value of the land and only in temporary employment for the landless. On the other hand, in such cases, if labour has to be hired at market wage, it immediately limits employment below what is socially desirable. In the Indian sub-continent such public works programmes have often been equated with *ad hoc* famine relief work outside the mainstream of activities of the national development plan. Often the necessary financial outlay is uncertain and beyond the resources available for public sector expenditures. Again, instead of involving the community, schemes are organised through labour-contractors.

Such programmes could, however, involve the local community through organisation of the landless labourers and the marginal or small farmers. Peasants could also organise themselves on a cooperative basis (as in some projects in Mexico, Sri Lanka and Bangladesh) to bring about the vital infrastructural development in their villages. This could also result in the participation of women in works programmes as has happened in Bangladesh.

Elements of these approaches to various aspects of rural development planning and implementation, including technology choices, industrialisation, public works, education and health policies and an enlarged role for women, are all to be found in many Third World countries today. They are, however, mostly applied in very marginal and partial ways, through small pilot-schemes that are usually implemented through the assistance of non-governmental, voluntary agencies. At the same time, similar approaches appear to be quite viable for the participation of the rural poor in development. Although many governments are unable or unwilling to undertake the basic nationwide institutional reforms essential for effective rural development, this does not preclude the *systematic search* for appropriate technical solutions and institutional forms through local experimentation.

These experimentations have to be identified, supported and stimulated, so that the rural poor can organise themselves. It is only through plural organisations at the local level i.e. the peasants' association, the women's association, the co-operatives, the trade-unions, landless labourers, the youth brigade, that more and more voices would be heard. It is only with the commitment of the majority population and under strong community surveillance that real development can move forward. Otherwise rural and national underdevelopment will continue to deteriorate.

REFERENCES

1. "*What Now*", the 1975 Dag Hammarskjold Report, Uppsala, 1975.
2. Address by R. MacNamara, President of the World Bank to the U.N. Economic and Social Council, 28, October 1971.
3. A.Z.M. Obaidullah Khan and Tomas Hexner. *A Turning Point : A Working Document for the A.C.C.*31 January 1976.
4. Khan and Hexner's Report, Ibid.
5. Mark Selden, *The Yenan Way*, pp. 93-94.
6. Abridged from *The Agrarian Prospect in India* Daniel Thorner, 1976, pp. 73-76.
7. Zubeda M. Ahmed, *Land Reform in Asia with Particular Reference to Pakistan, the Philippines and Thailand,* ILO, Geneva, 1977.

8. Memorandum from Secretary Conardo F. Estrella to all regional directors, 9 January 1973.
9. Samir Radwan, *Agrarian Reform and Rural Poverty in Egypt*, 1952-75 ILO, Geneva, 1977.
10. Ibid., pp. 41-45.
11. Hollis Chenery, et al., *Redistribution with Growth*, Oxford University Press, London 1974, "Rural Target Groups", pp. 113-135.
12. Gibert Etienne, *The Global Process of Rural Development in China in Studies and Documents*, Asian Documentation and Research Centre of Studies, Geneva.
13. Norman T. Uphoff and Milton J. Esman, *Local Organisation for Rural Development : Analysis of Asian Experience*, Rural Development Committee, Center for International Studies, Cornell University, Nov. 1974. Countries analyzed are Bangladesh, China, Egypt, India, Indonesia Israel, Japan, Malaysia, Pakistan, Philippines, South Korea, Sri Lanka, Taiwan, Thailand, Turkey and Yugoslavia.
14. Rural Development Expansion Programme in Bangladesh, Government of the People's Republic of Bangladesh, 1977.
15, 16. ILO, Geneva, 1977 op. cit.
17. Ibid.
18. John P. Lewis, "Note of a Rural Area Development Tourist", September 1973. (mimeographed).
19. Amisur Rahman, Adapted from *Participatory Organisations for Rural* Poor.
20. G. Tschannerl, "Rural Water-Supply in Tanzania : Is 'Politics' or 'Technique' in Command ?" Paper presented to the Annual Social Science Conference of the East African Universities, 1973.

THREE

PEOPLE'S PARTICIPATION
Focus on Mobilization of the Rural Poor

Ram P. Yadav

Introduction

THE ESSENTIAL theme of development has been the elimination of poverty, social inequality and unemployment. These have become national goals for each developing country. The strategy used to achieve these goals focused largely on economic growth in the belief that an increase in gross national product will percolate down to the poorest and thereby eliminate poverty, inequality and unemployment in a country. The planning model which performed miracles in the Western developed countries and the centrally planned economies were too readily adopted as ideal and appropriate to the developing countries. But after over a quarter of a century's experience and experiments with it, the growth model stands as the "god that failed". A rising growth rate has been no guarantee against poverty. To quote Mr. Mahbub Al Haq "we were taught to take care of our GNP, since this would take care of poverty. Let us reverse this and take care of poverty first, since GNP can take care of itself, for it is only a convenient summation, and not a motivation for human effort."[1]

Thus began in the early seventies a search for "Another Development".[2] Development began to be conceived to be more than economic growth. Equity from the point of view of narrowing the ever widening gap between the rich and the poor has been recognized to be an inalienable dimension of development.* It is also realised that unless the living condi-

* Irma Adelman, and C.T. Morris, *Economic Growth and Social Equity in Developing Countries*. Stanford University Press, Stanford 1973.

tions of the majority of rural poor who live below subsistence level, are improved there cannot be development in its true sense. This highlights the "basic minimum needs" approach to development.[3] Thus in recent years there has emerged, by and large a consensus that integrated rural development may offer an alternative strategy for development in these countries.

Rural development specialists have defined rural development in many ways but the gist of these definitions boils down to a primary goal of providing an opportunity for decent living to the mass of the low income population residing in rural areas on a self-sustaining basis. More specifically the objectives of integrated rural development are as follows :

1. Increase in production and productivity
2. Equity :

   (i) in access to opportunities to earn income,
   (ii) in access to public services,
   (iii) in access to productive inputs

3. Gainful employment
4. Self reliance
5. People's participation in development process
6. Ecological balance i.e. proper management of physical resources such as land, water and forests.

These objectives are inter-related and hence an integrated approach to development implies that any trade-off between or among these should be minimized.

Although each of the new integrated sets of objectives is important, this paper is devoted to an analysis of issues involved in "people's participation" which, in my opinion, is pivotal to the new strategy of development.

People's Participation as the Centre Piece in IRD Strategy

When people begin to participate in all aspects of the development process, namely, decision-making, implementa-

---

R.S. MacNamara, Address to the Board of Governors (Nairobi) Washington D C. International Bank for Reconstruction and Development 1973.

Hollis Chenery, et. al., ***Redistribution with Growth,*** Oxford University Press, London 1974.

tion, monitoring and evaluation and benefit sharing, other objectives of the new development strategy are also likely to be fulfilled. For instance, people's participation in planning and implementation of development programmes and projects leads to the selection of the types of projects which are of direct benefit to them and will also generate more gainful employment. The utilization of idle labour resources in productive employment will lead to increases in production since local resources i.e. labour, indigenous material and technology are used, the process is likely to become self-reliant.

The most important form of participation is in sharing of the benefits of development projects. This brings in the equity aspect of development. Similarly, the improved management of physical resources such as land, water and forests is possible only through people's participation. We now know that group or community decision is extremely important in the proper use and conservation of these physical resources. The evidence is based on the many successful cases of community irrigation projects, community forest or soil conservation projects. Thus we find that all other objectives of integrated rural development revolve around the pivot of people's participation which makes it the centre piece in the alternative strategy of development.

MEANING AND FORM OF PEOPLE'S PARTICIPATION

People's participation means involvement of the people in the development process *voluntarily* and *willingly*. Such participation cannot be coerced.

People's involvement has to be understood in the following four senses :

(i) participation in decision making;
(ii) participation in implementation of development programmes and projects;
(iii) participation in monitoring and evaluation of development programmes and projects; and
(iv) participation in sharing the benefits of development.

I would also like to limit the meaning of people's participation only to direct participation, and not to their indirect participation through their representatives. This is because, given the existing socio-economic political structure of the societies in most of the developing countries, the so-called "representatives" of the peoples are most likely to represent the rich rather than the interests of the poor majority. Since direct participation of the people is possible only at the local, village or community level, the rest of this paper will focus on people's participation in development at the local or village level. This is also in consonance with the assumptions made in the paper that an equitable sharing of the benefits of development by the poor is possible only if there is equitable participation by them in the process of development. True participation is possible only if the poor can mobilize and thereby gradually work their way upwards, so to say, in influencing higher institutions of political power with their united strength. This "bottom up" approach in target-group oriented participatory development is central to this paper.

A logical conclusion that can be drawn from these assumptions would be that given the existing social systems, equitable participation of the poor in the process of development can only be a gradual process. The process can be accelerated only if the poor become conscious of their rights and privileges and build up faith and confidence in themselves through united strength to achieve justice for themselves in the sharing of benefits of development.

People's participation in the four aspects of the development process is discussed below.

Participation in Decision-Making

Presently there is virtually no participation of the rural poor in any kind of decision-making process at the village, district and national levels of government. The normal procedure is that the guidelines for resource utilization, budgetary allocation to different sectors and regions, and the programmes and projects for area development are prepared by the central government. Even when such programmes are formulated in consultation with political representatives and corresponding budgetary appropriation is approved by legislative bodies at various levels, the interests of the poor masses are less likely to be given high priority since the political system, by and large, represents the elite groups and reflects the needs and interests of elitist minorities. Thus the participation of the poor indirectly through their so-called "representatives" is most unlikely to be in their interest under the socio-economic-political structures prevailing in most of the developing countries. Hence, participation by the poor is effective only if it is direct.

Direct participation of the rural poor in decision-making is only possible at village level. At the village level also, discussion regarding *community projects* such as *drinking water, school construction, feeder road construction, health post construction* and others should be made not in small caucuses but in open assemblies in the presence of all the villagers. In the beginning the rich are likely to influence the decisions in their favour in setting priorities of these projects and their locations. But gradually as the whole community gets involved in the decision making process it is unlikely that decisions would be grossly unfavourable to the interests of the poor. In course of time the poor will begin to participate more and more effectively and will be in a position to influence decisions in favour of the needs and interest of the majority.

The open forum of village assemblies as suggested above will be more conducive in raising important issues such as the following. When a project such as a road is constructed in an area, who should make the final decision in the alignment of the road ? Should not the alignment suggested by engineers presumably on technical grounds be discussed by the villagers and the engineers be asked to give justification, on a purely technical basis, to ensure their accountability to the people ? Shouldn't the community be involved in discussing the rate and method of compensation that is to be paid to those whose lands have been used in the road such as taxing those who would benefit directly from the windfall gain in the value of land due to road construction ? Discussion in open village assemblies of issues like this which cut across class interests, will also prod the villagers, particularly the majority, into participation in decision making in their own interest, while the "leaders" will also be put on alert to justify their position in public.

The open assembly, is one form to enable the poor and the majority to participate directly in decision making at the local level. It may not necessarily be the most effective means and is in danger of being manipulated by the vested interests to preserve their own power structure. The issues, therefore, that need to be further explored and discussed are as follows. Are there other forms whereby the participation of the poor in decision making for development can be made more effective ? What is the experience with innovative measures that may have been devised and tried in some of the developing countries ?

Participation in Implementation

The normal practice of people's participation in implementation of community projects has been in the form of contribution of free labour, voluntary or otherwise. It is the poor who possess only labour as their sole asset, and who are asked to contribute their only asset without any compensation. The rich on the other hand are not required to make proportionate sacrifices and it is they who benefit most from the community projects. Hence, this practice of people's participation through free labour is both inequitable to and exploitative of the poor.

An equitable participation in contribution for a community project is possible only when everybody in the village either voluntarily contributes or is required by the community to contribute an equitable share either in the form of free labour or cash in proportion to the benefits that each is expected to derive from the project when completed.

Another small but essential aspect of participation in implementation is that of the maintenance of community projects after they have been completed. The lack of maintenance of community projects has often reduced their lives and utility. Therefore, people's participation in project maintenance on a community basis is very important. How the community should be organised in an equitable manner for the maintenance of community projects is therefore a related and a very relevant issue that needs to be further explored and discussed.

The basic issue in participation of the poor in implementation is how to compensate them in an equitable fashion either in the form of payment for their labour or in the form of benefits to be equitably shared by them as members of the village community. Compensation for labour would raise the need for resources and simultaneously the issue of resource constraints. An innovative measure to meet this challenge on a much larger scale has been suggested in Mr. Ram C. Malhotra's paper.* Are there any other innovative measures being tried elsewhere ? What are the experiences with them ?

Participation in Monitoring and Evaluation of Programmes and Projects

Monitoring and evaluation are essential to provide feedback to the project management and policy-makers regarding any corrective actions to be taken or adjustment in the policies to be made during the implementation of a project. These aspects of the development process are very closely related to project identification, formulation, appraisal and selection aspects discussed under (1) namely participation in decision-making. The two aspects are also very similar in that, like participation in decision-making, there is also very little participation by the majority and the poor in monitoring and evaluation at present.

> Monitoring is the continuous gathering of information on project inputs and objectives and on conditions and complementary activities that are critical to the success of the project. On going evaluation is the continual analysis during project implementation of project outputs, effects and development impact. Ex-post evaluation is on analysis after the completion of the project of its (output), effects and impact.[4]

* Ram C. Malhotra "An Alternative Strategy for Self-Sustaining Development with Focus on Participation by the Poor at the Local Level" in Malhotra et al. (eds) *Alternative Strategies for Development with Focus on Local Level Planning and Development : Proceedings of a Consultative Meeting*. United Nations Asian and Pacific Development Institute, Bangkok, April 1979.

People's participation is necessary to provide information related to the progress of the project in order to identify problems and constraints in implementation. Their involvement in monitoring and evaluation is also essential to identify not only how many but also *who* benefits from a particular project. People's direct participation in monitoring and evaluation is also likely to reduce mismanagement of resources in a project. The people should also be brought to evaluate the performance of government personnel in projects being executed in their areas. This will make the government personnel accountable to them.

The participation of people in monitoring and evaluation will also have to take similar issues as their participation in decision making, discussed earlier, since the two are closely interrelated.

Participation in Benefit Sharing

This is the essential element that was left out in conventional growth oriented development strategies by assuming that the benefits of development would percolate to the majority. As a result, the issue of equity in distribution was neglected. Since people's participation in benefit sharing means equitable sharing of the benefits of development, it may be looked upon as the heart of the new development strategy. It is also the *sine qua non* of people's participation in development for "people participate most freely and fully in what is theirs".* Voluntary and willing participation is possible only if people benefit from such participation. Often people don't see the benefit directly and so lose interest in programmes and projects. They would identify projects as theirs only if they share the benefits of those projects.

The crux of the problem of people's participation in general therefore is to find viable institutions and techniques for ensuring equitable sharing of benefits of development by the poor themselves. This is the issue we turn to next.

* Agricultural Projects Services Centre; Report of Seminar-cum-Workshop on People's Participation in Rural Development in Nepal. "Experience with people's participation in other countries" by Mr. A.Z.M. Obaidullah Khan, August 1978, p. 20.

MOBILIZATION OF THE POOR

In an unequal society, a minority of the rich and powerful usually appropriates a large share of benefits for itself and leaves the majority of the people poor, absolutely dependent upon itself. The poor individually cannot stand up or speak out for their rights and privileges and remain meek spectators in the development process. Therefore, people's participation which implies effective participation of the majority is possible only if they are mobilized.

Mobilization of the poor, however, is not an easy task. In the face of their bitter experiences and economic hardships and social ostracization, the poor are often skeptical of and apathetic to any new proposals and measures ostensibly designed to improve their lot. On the other hand, there is the risk of the rich and the privileged being opposed to any efforts that are likely to threaten their power and prestige in the village community. Hence mobilization should be done tactfully and gradually in the initial stages to gain the confidence of the poor, increase possibilities of support from other sections of the community and forestall strong opposition by power-groups in the village.

A framework for the mobilization of the rural poor is discussed below. It is based on the experience of the small farmers development project in Nepal. The essential elements of the framework are as follows :

Benefit-generating Development Activities

Mobilization has to be around economic and social activities, participation in which gives the poor a hope of gaining more material benefits for themselves. Therefore, groups should be formed around productive activities.

Homogeneity of the Groups

Members of a group must be strictly homogeneous in socio-economic terms. Because people of the same economic and social strata have common problems and socio-economic conditions, they are likely to have common needs and interests. They should be able to talk freely among themselves about their problems and concerns. Even if the groups are functionally organized, such as poultry farmers' groups, the criterion of homogeneity in group membership would apply; otherwise such groups are likely to be dominated by a few rich and powerful members.

Membership and its Size

There should not be any pressure from outside with regard to membership in the formation of these groups. They should be left on their own to choose the members of their group. Thus only those who can work together well, will join together to form small groups. This will also bring about greater cohesiveness and durability of the groups. The membership in the group should be in the range of 10-25. Bigger groups should be discouraged as these are not conducive to the groups' cohesiveness.

Multifunctional Character

Groups need not be unifunctional. Rather, they should be multifunctional in scope. In other words group members should be involved in more than one productive activity. Small farmers, tenants and landless agricultural labourers are normally engaged in a mixed farming system where they grow food and cash crops, tend a few domestic animals and poultry and engage in small scale cottage industries or any other off-farm employment activities. Therefore in view of the diversified interests of its members, a group should be concerned with almost all the activities in which its members have interests.

Regular Meetings

Groups should hold regular meetings at least once in a fortnight to review progress of their ongoing activities and also to review their plans for the future.

Group Saving

A group saving scheme is an important aspect of group activities. Every member should be required to contribute in a saving fund regularly. The money saved should be used for advancing loans to needy members to meet their emergency and subsistence needs, thereby reducing their dependency on moneylenders or landlords. Resources of the saving fund may also be used for expanding group activities.

Group Organizer

People do not organize into groups themselves, unless they are talked to and convinced of the benefits to be derived from membership in groups. For this reason, a group organizer is required. Group organizers should be energetic and motivated to work with the disadvantaged and the poor. Group organizers must be willing to mix with the poor and to live with them. They must conduct socio-economic household surveys to identify people of similar status and interests for organizing them into small homogeneous groups.

Group Leader

A group leader is selected from the group members by consensus. Other members are also assigned different responsibilities. Thus everybody shares the work of the group.

Credit

As groups are organized around productive activities, the need for credit arises. Banking institutions should advance credit to members of small groups, not on the basis of collateral but on the basis of the collective guarantee of the group concerned. Alternatively, total land assets of all the members of a group may be used as collateral for loans to individual members for the purposes approved by the group concerned. Similarly, collective loans may also be secured for collective or joint projects to be undertaken by groups.

Associations

As several small groups are formed and as they are involved in several activities, it will be useful to share these experiences to learn from each others' mistakes and successes. The several small groups in a village or an area, therefore, may be federated into an association which can occasionally meet and discuss common problems of member groups. Such an association can also undertake, on behalf of their member groups, joint marketing of their products or procurement of inputs with a view to taking advantage of the economies of scale in production.

Small Farmers' Cooperative Society

The association can then grow into a cooperative society fully controlled by the small farmers' groups. The board of directors of the cooperative society may be elected by representatives of its member groups or may be composed of group leaders serving on the board for a fixed term on a rotational basis. The group organizer, referred to earlier, will be attached to the small farmers' cooperative society and his primary responsibility would be to assist in forming groups and helping them in preparing their group plans with the assistance of extension agents of different departments at the local or district level.

Group organizers and managers of the cooperative society should be provided by the banking system, at least for the initial formative period of 5 to 10 years or until the cooperative societies themselves are in a position to employ their own group organizer and cooperative manager. These officials should be held accountable to the general membership of the cooperative society and its constituent groups for their performance.

The formation of the association and its transformation into a cooperative society must not be hastened or forced in a structured or bureaucratic manner. The higher institutions should be allowed to gradually develop themselves as need for them may be perceived by the membership of the small groups. It is the smaller groups that are the most important and need to be nurtured very gently as they have so far proved to be the most viable institutional forms for mobilization of the poor, as evidenced by experience with small farmer development projects carried out by FAO/ASARRD in several countries in the Asian region. A brief description of the project in Nepal is given in the next section.

SMALL FARMERS' DEVELOPMENT PROJECT IN NEPAL

The small farmers' development project was started in March

1976 as a Field Action project by Agricultural Development Bank (ADB) with the assistance of ASARRD*/FAO. Two districts—one in the Tarai (Dhanusha), and another in the hills (Nuwakot)—in the Central Development Region of Nepal were selected for this programme.

The main objective of this project is to improve the living conditions of small farmers, tenants and landless agricultural labourers with an integrated programme of supervised credit through group work plans. Group Organizers/Action Research Fellows (GO/ARF) are assigned by the Agricultural Development Bank to the project. Their main task is to assist in organizing the groups and to help them in preparing their group production plans. In addition, they also conduct socio-economic household surveys in the project area to identify farmers who are of similar socio-economic status and interest and are willing to work together in some "nucleus" activities.

The group organizer meets small farmers individually or in groups and discusses the programmes of small farmer development with them in order to create interest among them to form small homogeneous groups. Only those farmers who have land less than 0.5 hectare are included in the programme. As the group is formed, the members discuss the productive activities that they are interested in and indicate the credit-needs for such activities. Each individual member's credit requirement is discussed by the group. Then a group action plan is prepared indicating the activities to be undertaken and the credit required for those activities by each member. This plan is discussed with the group organizer and is then submitted for loans to the local cooperative society which in turn receives its credit from the Agricultural Development Bank. A loan is sanctioned to an individual member on group guarantee basis, wherein the total land assets of all the members of the group are taken as a joint collateral by the cooperative society.

The members of the group meet once every month to discuss their individual as well as group activities. The secretary of the group presents up to date information on their savings accounts. Work responsibility is delegated to different members so that all feel a joint responsibility. Until

* Asian Survey for Agrarian Reform and Rural Development.

February 1978, the number of small farmers' groups formed in both the Dhanusha and Nuwakot districts was 72 with the involvement of 997 families. The total credit disbursement was US $ 133,500 and the total amount in group saving was US $ 2,030. The repayment rate was 92 per cent.

Initially the small farmers were reluctant to believe that they would get loans on a group basis for income-raising activities, such as rearing of milch buffaloes, sheep and goats, horticulture, cereal crops production, draft animals. Since it had never happened before that government officials had approached them and offered them credit they could not believe that it could happen now. The usual practice is for government personnel to go to the village head's house or the rich landlord's house. Visits by the GO/ARF to the small farmers' houses was unusual and naturally came as a surprise to most of the farmers.

In the beginning it was not easy to form groups because it took some time to create confidence in people. Also the big farmers/moneylenders began to threaten their clients that they would stop giving them credit even in cases of emergency if they joined those groups. Because of the continuous effort by GO/ARF, the small farmers slowly began to realise the value of the programme and formed groups for gaining access to credit for individual as well as group activities. Several groups have now started group orchards or group cottage industries, apart from numerous individual enterprises.

Other activities of the group are group saving schemes, compost making, improved sanitation, family planning and literacy campaigns. Each member of the group deposits five rupees ($0.40) every month to a fund called an Emergency-cum-Development Fund for group use. The fund is used to help individual members with medical or social expenses such as weddings and funerals, with a view to reducing their dependence upon the village moneylender. Money from this fund is also loaned to the needy members for consumption purposes whenever such necessity arises. This fund is also used to meet or supplement the expenses for enterprises undertaken jointly by the group members.

Members of the groups have constructed pit latrines in their backyard and many of them have adopted family planning measures. Thus step by step they are taking up more activities and many of these are financed by themselves. This approach thus leads to greater and mutually supportive self-reliance on the part of both the individual members of a group as well as the group as a whole.

Four essential elements for the success of this programme may be identified :

Homogenous Group

It is important that the members of a group belong to similar economic and social strata. In a heterogenous group, the rich few will start dominating decisions in the group. A good example is provided by the traditional cooperative societies which have their membership drawn from different socio-economic strata and their activities largely dominated by a few large farmers who serve their own interests. The organization of the small farmers into cohesive and viable groups enhances cooperation for their common interests and improves their collective bargaining power. Therefore formation of groups from the same socio-economic strata becomes a necessary pre-condition for the success of this programme.

Credit

Credit is the key to this programme. It provides the main incentive for individual small farmers to join a group, because they can get credit only through group recommendation. The group is held jointly responsible for repayment of loans. Group discipline is also maintained through this credit scheme and each individual in the group has to go along with the majority decision because disobeying the group decision may mean expulsion from the group. This not only means that he will have difficulty getting loans but also that other groups will be reluctant to accept him as a member.

Support from Government and Related Agencies

The various technical departments and agencies of government should orient their activities to the needs of the groups. An efficient delivery system is crucial to meet the demands coming from these groups, otherwise the whole programme will receive a set-back.

GO/ARF

The GO/ARF is seen as a new type of community organizer. Initially, the success of the programme depended largely on his initiative and motivation. He must have a missionary zeal and be a person really concerned for the welfare of the disadvantaged section of the society.

Thus the programmes for the rural poor should have four essential elements to make any substantial impact on their living condition.

1. Increasing the income levels of the poor by providing access to more economic opportunities.
2. Reducing the dependence of the poor on the rich through some kind of saving schemes.
3. Promoting unity among the poor by organizing them into groups and encouraging them to take up economic activities jointly such as group farming, or group orchards and group marketing.
4. Increasing their consciousness and awareness of their rights and duties in society through education and informal meetings.

The small farmers' development programme in Nepal has covered all these four aspects simultaneously.

It has been observed, however, that in areas where the degrees of socio-economic disparity between the rich farmer and the poor farmer is great, this programme encounters more hurdles from the rural elite, since they feel threatened by the programme. The small farmers' development programme has encountered less opposition or hurdles in the hills where the disparity is relatively less as compared to the Tarai where the disparity is greater. An examination of the relationship between the degree of economic disparity in an area and the effectiveness of programmes for social change is interesting but beyond the scope of this paper.

REFERENCES

1. Mahbub Al Haq, *The Poverty Curtain*, Columbia University Press : 1976, p. 43.
2. The 1975 Dag Hammarskjold Report, "What Now : Another Development."
3. ILO, *Employment, Growth and Basic Needs : A One World Problem.* Geneva, 1976.
4. Technical Workshop on Monitoring and Evaluation of Rural Development Projects and Programmes, December 6-10. 1976, Copenhagen. Summary Report, P. 2.

SUGGESTED READINGS

Adelman, Irma and C.T. Morris, *Economic Growth and Social Equity in Developing Countries*, Stanford University Press, Stanford, 1973.

Agricultural Projects Services Centre, "Report of Seminar-cum-Workshop on People's Participation in Rural Development in Nepal", July 24-28, 1978 Kathmandu, Nepal.

Chenery, Hollis et al., *Redistribution with Growth*, Oxford University Press, London, 1974.

Cohen, John and Norman Uphoff, "Rural Development Participation : Concepts for Measuring Participation in Project Design, Implementation and Evaluation", Rural Development Committee, Cornell University, 1976.

Dag Hammarskjold Foundation Report, "What Now : Another Development", 1975.

Development Alternatives Inc., "Strategies for Small Farmer Development : An Empirical Study of Rural Development Projects", Washington D C. 1975.

FAO, "Learning from China" : A Report on Agriculture and the Chinese People's Communes by an FAO study mission, 9 September-5 October 1975. Bangkok 1977.

FAO, Report on the FAO/SIDA/DSE Inter-regional Symposium on Integrated Rural Development held in Berlin., Reiherwerder 19-23 September 1977.

FAO, Papers presented at the National Workshop on Two-way Process in Planning, Trisuli, Nepal, 20-24 March 1978.

Government of India, Planning Commission, "Report of the Working Group on Block Level Planning", 1978.

Haq Mahbub Al, *The Poverty Curtain*, Columbia Univ. Press 1976.

Haque, Wahidul, et al., "Towards a Theory of Rural Development", *Development Dialogue*, 1977 : 2.

International Labour Organization, *Employment, Growth and Basic Needs : A One World Problem*, Geneva, 1976.

Lele, Uma, *Designing Rural Development Programmes : Lessons from Past*

Experience in Africa, Johns Hopkins University Press, 1975.

MacNamara, R.S., Address to the Board of Governors (Nairobi), Washington, D.C. International Bank for Reconstruction and Development 1973.

Marga Institute. "Local Level Planning with the Framework of Overall National Planning," Sri Lanka, June 1976.

Obaidullah Khan, A.Z.M. & J. Tomas Hexner, "Poverty-Oriented Rural Development and the UN System : A Turning Point," A Working Document for the ACC., 31 January 1976.

Ruttana, Vernan W. "Integrated Rural Development Programmes : A Skeptical Perspective", *International Development Review*, 1975. 17(4) : 9-16.

UNDP/FAO; Asian Survey of Agrarian Reform and Rural Development "Starting from Below" Regional Project Findings and Recommendations, 1977.

Uphoff, Norman T. and Milton J. Esman, *Local Organization for Rural Development : Analysis of Asian Experience*. Centre for International Studies, Rural Development Committee, Cornell University, 1974.

Waterston, Albert, "A Viable Model for Rural Development" in *Development Digest*, Washington, D.C. July 1975.

FOUR

BASIC FACTORS INHIBITING "PLANNING FROM BELOW" AND SUGGESTIONS FOR OVERCOMING THEM

G.C. Clarke

Background

DEVELOPMENT literature from Asia over the past 30 years has not been hesitant in drawing attention to the importance of people's involvement in planning. Although not usually specified, it was generally intended that the term "people" referred to all people, irrespective of social or economic position. Yet, for some reason, practice has fallen far short of theory.

It was against this background that the member government of FAO, during their Eleventh Regional Conference in New Delhi in 1972, requested the Regional Office of FAO to look into the institutional constraints inhibiting the participation of the smaller farmers, including the landless agricultural workers, in agricultural and rural development Thus. over the past six years, FAO has been giving special attention to the problems of the small farmers, tenants, sharecroppers and landless agricultural workers in Asia.

Eight member governments—Bangladesh, India, Indonesia, Korea, Nepal, Philippines, Sri Lanka and Thailand—actively participated with FAO in this exercise conducted under a FAO/UNDP Regional Project entitled "Asian Survey of Agrarian Reform and Rural Development," now commonly known as ASARRD. Through a problem-identification, prolem-solving process of 16 field level workshops, 8 round-up sessions in the national capitals and one inter-country workshop in Bangkok involving over 1,500 officials and farmers from these countries, there emerged a set of "essential elements" for small farmers' development. By this was meant that any programme or project which wanted to involve and benefit the bottom majority, should make provision for the following six "elements" in its design :

(1) Field workshop (multi-level, problem-solving)
(2) Small grass-root groups
(3) Group organizers
(4) Special programme including reservation of inputs
(5) Micro-level "planning from below"
(6) Action-based research

Thus, once again, there was general agreement reached that "planning from below" was not only desirable, but essential. While this was far from an original finding, the methodology used in its rediscovery by such a large and widely variable group may have special relevance to this expert meeting.

The Basic Factors

The experience of ASARRD tended to confirm the presence of three basic factors inhibiting government planners and national development officials alike from seeking greater involvement of field personnel and rural people in the planning process. We will briefly discuss each of these factors.

(1) There is an unwillingness on the part of most senior officials to recognize that their subordinates, especially their field staff, and the rural people, including the rural poor, have knowledge and insights without which no programme or project can be planned or implemented. Basically they are unwilling to delegate responsibility, particularly in so apparently important a matter as planning. This appears to be due to a lack of trust in their subordinates and a fear or, at best, a reluctance to admit that there are things they do not know. Closely related to this is the reluctance to recognize the need for inter-departmental and inter-ministerial cooperation in tackling rural development.

(2) There is also an unwillingness among most officials to recognize that rural development in general and agricultural development in particular, is location specific and therefore requires a high degree of local decision-making. It is only through releasing the creativity and initiative of individuals and small groups which make up rural society, including the millions of rural poor who form the bottom majority, that rural development in Asia can take place. Instead of worrying about how to secure their active participation and thereby their initiative through involving them in the planning process, most officials still cling to the hope that rural development, like a factory, can be made to work if only they could better refine their present process of planning from the top.

(3) The third basic factor inhibiting a wider use of "planning from below" in the planning process is the absence of an effective mechanism for consulting with the rural people, particularly the bottom majority. Senior officials of government in general have failed to recognize that deliberate efforts must be made to create this mechanism, especially if the voices of the small farmers and the landless are to be heard.

In summary, it may be said that "planning from below" will only be practised when those now responsible for planning make "three recognitions"—

(a) *recognize* that the rural majority, including field staff have knowledge and insights essential to plan formulation and implementation and can be relied upon to perform well if given the responsibility and authority.

(b) *recognize* that the initiative of the masses, so essential to plan implementation, is most effectively secured and assured through their involvement in the planning process.

(c) *recognize* that deliberate and special efforts must be made to create a mechanism for "dialoguing" with the rural people, especially the rural poor who form the vast majority in most of the countries of Asia.

FAO's Strategy

The Regional Office of FAO saw the problem as largely one of attitudes. Therefore the methodology employed by FAO under the ASARRD project was selected deliberately to help develop an awareness amongst governments and government officials in the Region of the need to acknowledge the above "three recognitions."

We looked for a method of conducting projects which would demonstrate to the senior participants that their field staff and the rural poor had knowledge and insights about their local situations which were indispensable when designing projects or programmes aimed at helping that target group. To achieve this, we adopted three approaches. The first was the multi-disciplinary, multi-level, problem-identification, solution-seeking field workshop approach which was selected[1]; the second, a field action project approach and the third approach focused on developing a "receiving/utilizing mechanism."

The Field Workshop

Governments were requested to identify two or three of their lowest level administrative units for analysis. All government departments and agencies having programmes intended for the small farmers and landless agricultural labourers in those administrative units were asked to send representatives from each of their different administrative levels. The objective was to have representatives from each level of each agency to come down to the field and to see at first hand how their programme was reaching the villages and who was benefiting from it. In reality, it provided an opportunity for policy makers, planners and implementors from different levels of each department and agency to "view from below," the working of their organization. And by having all agencies involved together, it was hoped that the need for closer interdepartmental cooperation would become self-evident. We felt this to be important as a first step in convincing line departments of the need for integrated rural development. Since the review was to be based on how the rural poor saw these programmes, it was only logical that at least 20 per cent of the participants in each workshop were actual small farmers, tenants, share-croppers and landless agricultural labourers.

Each field workshop lasted about five days and was held as close to the area under analysis as possible. The participants totalled between 80 and 100 persons. Participation was multi-disciplinary, multi-departmental, and multi-level. About 20 per cent came from the national level, 60 per cent from provincial, district and field levels and 20 per cent from the rural poor. To avoid inter-departmental rivalry and any tendency to blame each other for lack of cooperation, the workshops focused on the specific field problems as identified first by the small farmers and secondly by the field workers of the different line departments. The first two days of each workshop were devoted to the process of problem identification.

The Process of Problem Identification

This process was undertaken with specific learning objectives in mind and included the following steps :

(1) presentation of an analysis of the problems of the area as seen by a recognized research body, followed by discussion;

(2) preparation of lists of the achievements and problems of the area as seen, firstly, by the participating small farmers and secondly, by the participating field workers of the different departments. This was done by dividing the participants into small, multi-level, multi-departmental groups of 8-10 persons. Focus was placed on these two categories of persons as they have been most frequently ignored in the planning process and unless it became evident in the process that their contributions were necessary, there would be no need to consider their inclusion in any revised planning procedure.

At the end of the first day the groups compared their lists of achievements and problems. While many groups listed similar problems there were always those who identified new and often basic problems. This impressed upon the participants the value of group action and multi-disciplinary and multi-level discussions in identifying new problems and developing new insights into old problems;

(3) visits, during the second day, to nearby villages where participants were divided into smaller teams of 2-3 persons for family consultations in the homes of one small owner-farmer, one tenant or share-cropper and one landless agricultural labourer. This provided participants an opportunity to see and "feel" at first hand the problems of this target group and of equal importance, to compare their findings with the list of problems developed by the discussion groups during the first day. This comparison is most important. If a senior official finds that the problems he insisted on having listed the first day, are in fact not the same as those he finds during his family consultations, he suddenly "learns" that he is lacking in some knowledge and therefore needs "planning from below."

(4) revision of the list of achievements and problems of individual discussion groups and preparation of a consolidated list acceptable to all groups. This included a regrouping of the problems into major problem areas such as land problems, credit problems, extension problems etc. This helps to systematize the next stage of solution-seeking.

One of the important lessons learned from the above process was the need for planners to balance the general findings of research institutions with personal fact-finding at family and village level. The personal experience of direct dialogue with the rural poor proved indispensable for correct analysis and interpretation of the general research findings and led to more realistic plans.

The last three days of each workshop were devoted to seeking practical solutions to the commonly agreed upon problems and bottlenecks. The task was not as simple as initially thought, especially by the senior officials. The exercise proved to be humbling for all concerned and served to develop amongst those from national and provincial levels a greater awareness of the vital knowledge and insights which district and field-level staff and practising farmers and agricultural labourers have to bring to the process of planning for successful implementation of specific agricultural and rural development programmes.

In summary, the Field Workshop approach has proven to be a useful vehicle for :

(1) developing positive attitudes towards planning from below amongst senior officials;
(2) helping governments review their ongoing rural development programmes in a manageable manner, leading to programme and policy changes more favourable to the small man; and
(3) demonstrating the need for multi-disciplinary and multi-departmental planning in order to achieve a more integrated approach to rural development.

Field Action Projects

The second approach under FAO's strategy to promote "planning from below" is through encouraging governments to launch field action projects (FAPs)[2] for small farmers' development based upon the "essential elements" identified under the ASARRD project.

There are presently two FAPs in Nepal, three in Bangladesh and three in the Philippines involving over 3.000 families of the rural poor. Focus is on helping them organize themselves into small, homogeneous groups of 10-15 people for income-generating purposes. Banks are extending credit to these groups under joint liability. Over 260 such groups have been formed over the past two years. Most of them are functioning surprisingly well with loan repayment above 90 per cent in all three countries. A full-time group of 18 Group Organizers guides this critical process and they also record and analyse the results as they happen. Thus the project is one of action research.

None of the FAPs have completed their full three-year research period. However, early findings tend to attribute their present success to the fact that :

(a) membership in each group is determined by consensus of all members and that (b) they select and plan their own group economic activities. This is group planning from below in its purest sense and clearly demonstrates the importance of the "second recognition"—that of the need to release the peoples' initiative from the outset through their involvement in planning.

Developing a "Receiving/Utilizing Mechanism"

The third element of FAO's strategy to promote a better understanding of the need for "planning from below" is to stress the necessity for planners to make a clear distinction between efforts to improve the "delivery mechanism" of line departments and efforts to develop a "receiving/utilizing mechanism" for the small farmers. The latter is necessary in order to take advantage of the inputs and services being made available by the line departments and agencies of government. Without such a mechanism for the rural poor to help themselves, they will continue to remain outside the main stream of development. Thus, the basic importance of helping members of this disadvantaged sector to organize themselves into small production-oriented groups. Through group action and greater economy of scale they are, at least, able to move effectively into the competitive society which characterizes the economic systems of most countries in this region.

It is ironical that while these small groups are proving to be indispensable in making their own membership more self-reliant they are, at the same time, proving to be equally essential to governments as a vehicle for determining the real interests and needs of this sector for effective planning.

FAO is encouraging these groups to undertake a series of planning steps. The first is planning for their joint income-generating activities. From this exercise, the group is being encouraged to help each of its members develop his or her own farm/family plan. This is the second planning step. The third step is to encourage the different small groups in one or more villages to come together to plan activities of common interest. This leads to establishing their own associations, as is now being undertaken, in a limited way, by some of the FAP groups in Bangladesh, Nepal and the Philippines. The fourth step then is to match these association plans with those being developed by government departments and agencies at higher levels. It is at this point that "planning from below" and "planning from the top" can merge—where each can learn from the other and contribute to the development of, national, regional, district and local plans which combine national and local interests, including the special interests of the rural poor The entire process cannot be started without the organization of the grass-roots groups.

The potential of this approach is yet to be fully explored. But the results of an extensive National Workshop on the Two-Way Process in Planning held in Nepal in March, 1978 were very positive. At this workshop, held in a remote district far from Kathmandu, members of the National Planning Commission and planners from the Ministry of Agriculture and the Agricultural Development Bank of Nepal met with about 20 small farmers' group members from 10 different small farmers' groups from both the Tarai and the Hills. The conclusions were positive. The importance of these groups as a "receiving/utilizing and planning mechanism" in any two-way planning process was fully recognized. Thus the small farmer groups demonstrated to the national planners their importance in any effort to promote increased "planning from below".

Conclusions

The manpower deployment situation prevailing in many countries of this region is a relatively small number of highly trained, experienced yet grossly over-worked officials at the national level, with a large layer of well trained, but highly under-utilized officials and technicians at middle and field-levels waiting for instructions from the top. Local initiative and local decision-making is not encouraged. The opposite is required for rural development to take place.

A trend towards greater decentralization of both administration and planning will only come about when those with power at the national level feel they can entrust matters to those below them. A programme of location-specific field workshops as described earlier in this paper has proven to be a useful educational tool for letting such top level officials "experience" the valuable contributions which lower level officials, technicians and the small farmers themselves can make to more realistic planning. Those who participated claimed they found the exercise very rewarding and now realized the importance of local participation in planning. Thus it may be claimed that attitudes were changed. Positive learning took place.

In addition, field workshops teach the value of multi-level participation and the importance of multi-departmental problem-solving as a basis for realistic rural development planning. A wide variety of objectives can therefore be reached through the single method of conducting problem identification, problem-solving-cum-planning field workshops. It may be argued that every official should have the opportunity, at least once a year, to participate in such a bottom-up planning exercise, both as a means of keeping in touch with real field problems and as a means of reinforcing one's faith in the wisdom of the masses.

Similarly, the development of several field action projects for small farmers' development have proven to be useful laboratories for testing new approaches to planning and new methods for securing the popular participation of the rural majority in rural development. The formation of grass-root groups for income-generating purposes amongst the rural poor has at the same time shown the way towards establishing a mechanism so essential for initiating effective planning from below.

The above measures would therefore appear highly useful, if not essential if one believes that rural development in general and agricultural development in particular, requires planning with, rather than planning for the rural majority.

REFERENCES

1. For details see "Small Farmers' Development Manual Vol. II : The Field Workshop : A Methodology for Planning, Training and Evaluation of Programmes for Small Farmers/Fishermen and Landless Agri-Bangkok, Labourers." FAO Regional Office for Asia and the Far East, cultural Thailand (78 pages).
2. For details see "Small Farmers' Development Manual Vol. 1 : Field Action for Small Farmers, Small Fishermen and Peasants". FAO Regional Office for Asia and the Far East, Bangkok, Thailand (206 pages).

FIVE

PARTICIPATORY DEMOCRACY, INSTITUTION BUILDING AND SOCIAL DEVELOPMENT

Salima Omer

ONE of the most significant challenges before Third World countries is that of achieving social, political and economic development along with social justice. An attempt is being made in this paper to present a theory of achieving social-political development through the active participation, involvement and control (PIC) of people in the development process. The hypothesis is that within the socio-political context of developing countries the participation, involvement and control (PIC) of people can give rise to a socio-political system more conducive to the attainment of economic development along with a greater degree of social justice.

The paper also presents a conceptual framework of institution building which the author believes facilitates the process of people's participation, involvement and control. It is contended that the existing socio-economic and political institutions do not provide the people an opportunity for active participation. Alternative decentralized institutions that operate at the level at which people are, could become instruments to foster participation and control of people at the grassroot level. An attempt has also been made to define social development which in the last analysis should be the goal of all economic growth.

Introduction of Political Systems and their Aftermath

We start this paper by briefly reviewing the recent political experiences of a few South Asian countries.

On the fifth of July 1977, the army in Pakistan took control of the affairs of the state, placing the Prime Minister, some cabinet members and a number of the members of the opposition, the Pakistan National Alliance (PNA), into "safe custody". This means that in a short span of thirty years, the military has taken over in Pakistan for the third time and the so-called normal democratic process has been interfered with. India, considered the bulwark of democracy in Asia, is also beset with many problems that lead us to believe that the parliamentary democratic process has still a long way to go. Sirsikar states that "the three general elections (of India) and the successful working of the parliamentary institution are by themselves no guarantee of a democratic society".[1] For nineteen months, the Prime Minister, Indira Gandhi, declared a state of emergency. During this period the jails were filled with leaders of the opposition, the press was gagged and all political activities were suspended. However, the very fact that the Janata Party came into power through the electoral process is a promising feature. During the period the Janata government took over, India's problems did not seem to have abated, and this leads us to believe that economic democracy is a commodity too expensive to buy even when success is achieved at the polls.

In Bangladesh, on second October 1977, some junior officers of the army attempted a coup to dislodge the military regime. The military regime itself had usurped power from the so-called political government which was formed after the founder of the nation of Bangladesh, Shaikh Mujibur Rehman was assassinated in a military coup.

Most post-war nations have professed belief in a democratic system of government. Mehden states that "definitions of democracy and timing are different, but the ideologies of almost all countries proclaim some sort of democratic system as the ultimate goal".[2] Attempts of other developing nations of Asia, Africa and Latin America to achieve a parliamentary or presidential system of democracy have met with similar scant success.

The reason is that many countries after gaining their independence from the colonial powers tried to introduce an electoral system based on universal adult franchise, the provision of political parties and so on. In other words, they tried to make provisions for the formal trappings of a representative democratic system similar to what obtained in the former colonial countries. However, they seemed to be oblivious to the fact that democracy is more than a form of government. It entails: (1) certain attitudes; (2) some degree of political loyalty, and belief in the worth and dignity of all people; (3) tolerance and a genuine effort at understanding the "other point of view"; (4) respect for the views of the opposition and an understanding of its role in providing certain checks and balances. The manner in which the representative system of the West operates also requires an educated and well informed electorate, a relatively free press and other media to reflect and mould public opinion. Briefly stated, democracy, besides being a way of life, requires institutional forms where such variables "as the degree of popular participation in the political process, the degree of freedom of the political opposition and the press, the degree of competitiveness among political parties and the degree to which the individual citizen regards it possible to have his input on public policy, are observed and roughly measured".[3]

In most developing countries these prerequisites did not exist but those at the helm of affairs were oblivious of the significance of these prerequisites and were keen to thrust upon their countries a foreign system of government and alien political institutions.

If a cursory analysis of the history of the last fifty to sixty years of the developing nations is made, one finds that country after country, the same story is repeated and the same drama is enacted, particularly in countries that were under the colonial rule. Simply stated, the drama presented in three acts is somewhat as follows, with some differences in degree, depending on the stage of development of a particular country.

Act I: A colonial power subjugates a country and replaces the indigenous social-political-economic system with its own which deprives the people of almost all forms of self-government. The entire nation and all activities therein are controlled either by the colonial power themselves or by indigenous officers carefully prepared to play the tune of the piper. The educational system is not geared to produce people with initiative and creativity but men who can carry on the day to day tasks of the government in power. H. Halsey comments as follows :

> In India before independence, 12000 Britons administered a country of 400 million people. This necessitated the British educational policy which was stated by Macaulay in 1835 thus, "we must do our best to form a class who may be interpreters between us and the millions we govern, a class of people Indian in blood and colour, but English in taste, in opinions, in morals and intellect".[4]

Besides, the system is geared to providing education to the rich and urban few. The people living in the rural areas, and particularly the poor, are kept out of the system as education is an expensive commodity that they cannot buy. The consequence of such a policy is that on the one hand, the indigenous system is distorted if not totally destroyed, and on the other hand, it is not replaced with another system capable of providing education to the population at large.

In the economic arena, the indigenous population carried out its daily activities around farming, shopkeeping, small businesses and continued paying taxes to the government in power. The colonial powers shipped the raw materials out of the country, processed and manufactured these and brought them in as finished products to an almost captive market. No efforts were made to set up industries at home to process the raw materials and thus to develop the colonial economies. The development of infrastructure, namely roads, railways, power and water, etc., depended on whether the foreign government needed these services to augment its revenues and to maintain law and order.

The primary objective of the colonial powers was to maintain law and order, and continue to carry away the spoils from the countries under their subjugation. Provision of education and other social services were to help provide a climate conducive to carrying out these primary activities, and hence offered at a minimum level. The motivation for any reform or services was the extent to which it was beneficial to the colonial power rather than whether it was in the interest of the people at large.

Act II : The scene changes to some extent when the indigenous populations agitate and demand a voice in the government. The violent and non-violent activities, bring about certain concessions for the indigenous population and some rudimentary form of self-government. It starts with local government and usually goes little further than membership in advisory councils to the governor or as nominated/elected members in state governments. At last the yoke of the foreign government is thrown away, independence is achieved and the third act begins.

Act III : The political party that has been instrumental in leading the country to independence forms the government and makes an effort to introduce a political system (usually the one followed by their former colonial masters). As time goes by the party in power, by and large, abuses its power, and seeks more to control the society than to provide opportunities for people to participate in the democratic process. Nepotism and unscrupulous activities create discontent and give rise to more political parties competing with each other for power and control. Such bickerings and, at times, violent differences shift the attention of the government from nation building activities to that of suppressing all political opposition and controlling the various dissenting groups. The effects on the economy of such turmoil are devastating. Local businesses and day-to-day activities are affected, export earnings reduced, the investment climate worsens and a loss of confidence in the prevailing government follows. Thus the political, economic, and social state of affairs frequently leads to military intervention and extra-constitutional government.

Briefly stated the third act portrays the political parties attempting to introduce a representative form of government, followed by the growth of more political groups vying with each other for power and control. The next step in this process is that of non-cooperation, riots and sometimes a civil war which sets the stage for a military take-over. In country after country, the same story is repeated, albeit with some differences in degree and sometimes in sequence.

An analvsis of this phenomenon leads one to five basic conclusions.

The first is that the elites of the country are making the most of the political and economic decisions and are devising a system of government on the basis of their own knowledge and preferences. Emerson describes this phenomenon as follows :

> Although the analogy is in many respects unfair and should certainly not be pressed too far, it may be suggested that just as the colonial governments were imposed on the people from above and outside, so the new Western style constitutions have been the work of small groups whose intellectual and, in a sense, political links are with the West rather than with the mass of their own people.[5]

Second, the systems chosen in the political and the economic arena are those that their former colonial masters namely, the capitalistic model in the economic arena and the parliamentary or the presidential model in the political system. Third, as regards the socio-political and economic development of the country, the population at large is not involved in the decision-making processes. Fourth, the models adopted by the elites are foreign and alien to the people and impose a lifestyle and perhaps even a belief system that is in conflict with the values that they hold sacred. Fifth, the prerequisites for the successful operation of models of the West are totally missing in the developing countries or are present only in a very rudimentary way. Thus the attempts of many developing countries to achieve political stability have met with scant success. Rather, the educational, political and economic systems introduced by the elites have aggravated the dualism present within the societies and increased the gulf between the haves and the have-nots. Even the agriculture policy aimed at achieving self-sufficiency in food, namely the green revolution, has heralded the red revolution and led to the development of the rich farmers and the further impoverishment of the small landholder.

Why Grassroots Participation

In the above analysis, we have tried to show that the political and economic elites have not been able to introduce a viable political system, nor economic growth with social justice. Their models—political, economic, educational—are alien to the people and do not reflect the people's values and aspirations. Brown, commenting on political parties as media for reaching the people, makes the following observation :

> The conversion of Asian, African and Latin American political parties into mass organizations (was) not simply a technical problem involving such matters as the creation of local branches and publication of party newspapers. Nor (did) it rest upon the triumph of the principle of universal suffrage. The real difficulty (was) that an enormous gap (existed) between a small, educated elite and the illiterate peasant masses. Not only (were) there few organizational links between the elite and the masses, there (was) hardly any communication.[6]

Another political scientist describes this state of affairs as follows :

> The juxtaposition of an elite educated in a tradition of exogenous inspiration and a mass rooted in a variety of indigenous cultures engenders problems that turn up throughout the world of the new states.[7]

In this paper we present the theory that the socio-political and economic development of the Third World nations could be better achieved through the participation, involvement and control of the people. Such an effort will give rise to a political system that is, at least in relative terms if not in an absolute sense, more conducive to the attainment of socio-political-economic development with greater social justice.

One of the problems of constitutional government and democracy in Asian countries is the fact that in these countries political changes have preceded social and economic change.[8] We shall attempt to show that to provide people a climate conducive for participation, involvement and control (PIC) it is necessary to resort to the strategy of institution building. We shall shortly enunciate what we mean by institution building, its scope and parameters.

Since the elites have not been successful in introducing and maintaining a political system capable of developing the country and its people, it seems only logical to look for another set of people who could deliver the goods. However, expediency alone is not the reason for seeking out the "people per se" to don the mantle of leadership. People have the inherent right to be involved in the decisions that affect their lives and representative democracy as pursued by the elites of the developing countries has not been able to provide the vehicle for such involvement. Another reason given for the non-involvement of the people in political and economic decision-making is that they do not possess the requisite knowledge and expertise to participate in the complex decisions of twentieth century problems. The facts, however, state very eloquently that it is the educated elites, often immersed in twentieth century scholarship, who have not been able to introduce a political system that could reflect the aspirations of the people and lead them to a better life. The apparent lack of education and even literacy is no deterrent to the involvement of the people in the decision-making-processes provided the decision-making processes are brought to their level and involve issues and problems that they are concerned with. The apparent lack of education does not mean that people do not possess intelligence and the wisdom to be able to discern programmes and activities that would help them to achieve a better life. Rather the process of people's participation, involvement and control (PIC) will lead to the determination of what constitutes a "better life" by the people themselves, rather than by some expert, local or foreign, who makes these decisions in the confines of the planning commission. The decisions as to what constitutes a better life would then also reflect the value system of the people rather than being at cross purposes with what people consider significant and sacred. "It is a sound principle of human order that social tasks should be left at the simplest and most human level at which they can be adequately performed—beginning with the family".[9] Thus it is possible that the process of involvement of the people would lead to social development that would make the process of political development easier and less hazardous.

Social Development

We define social development as a process and goal that aims at the total development of people in a manner that they deem fit and desirable. In other words, the people determine what is development, its parameters, and the trade-offs they wish to make for achieving a certain kind of development. Development is also considered as a total process, not a fragmented, piecemeal and isolated approach to issues and problems. Social development is inter-sectoral, inter-regional and inter-disciplinary and visualizes institutional and structural reforms to provide greater social justice. Social development encompasses within its domain strategies that aim at the total enhancement of the life of all people; its conceptual base is belief in the dignity of human beings which has as its logical corollary, the equality of all people. In other words, the concept of exploitation of some people of some region by people who have greater power and resources is alien to the concept of social development.

Decentralized Institution Building

With this brief digression on what is social development, we revert to our discussion of participation involvement and control of people in those socio-political processes that affect their lives. To facilitate this process the strategy of decentralized institution building could be adopted.

By institution building we mean the formation of mechanisms that afford a continuous and sustained forum for people to discuss the issues and problems that concern them, to decide ways and means to meet such problems and to plan activities that reflect their needs and aspirations. Such institutions could be as simple as that of a neighbourhood committee or as complex as a cooperative bank. They could be formal or informal, though it is possible that formalization may occur as these institutions strive to enhance and strengthen their work.

By decentralization is meant that such institutions would operate at levels very close to the people. The attempt would be to make the participation as direct as possible. Implied in the concept of decentralization is the belief that human activities, programmes and institutions can be so devised that it is possible to delineate its various parts and decentralize the planning and implementation of programmes. The smaller the unit of decentralization the more direct the participation and the greater the involvement and control of those directly affected. This process would also make the people more accountable to one another and thereby to the society at large.

Certain programmes and activities however, do need to be organized on a somewhat larger scale. If institutions need to be organized at relatively higher levels the process of representation could be used but this process should not be far removed from its geographical constituency. We are aware that small units cannot take over the multifarious work that is required to meet community needs and the needs of the larger society. To accomplish such work and to resolve issues that require the cooperation of more people and larger bodies it is possible for two or more units to join hands or engage in a more formal federation to meet their mutual needs or to tackle their multifarious problems. For example, many neighbourhood committees could form themselves into a council to set up a secondary school or to move the government to establish a secondary school that could serve their neighbourhood. We can build decentralized institutions which involve the setting up of institutions at various levels to attend to tasks that require the efforts of several grassroot organizations. However, the institutions at the various levels would be so organized that their composition would reflect as directly as possible representation of people from one echelon to the other.

Decentralized institutions in the Third World countries if properly organized, can, over a period of time, achieve the following. They can :

1. *Create more leadership* of various kinds and at a number of levels. For example, neighbourhood groups and councils at higher levels, would provide greater opportunities for people to participate in meeting the needs of the local population, and in utilizing their skills for the general improvement of the local level.
2. *Give people a greater voice in decisions* that affect their immediate environment—social, economic and political—as well as in matters that affect their surrounding communities. As the institutions at the local levels and at higher echelons develop, the participation of people would become more intense and lead to local control over decisions and resources. As the number of levels of institutions increase, people's participation at various levels will also increase giving them greater control of their own affairs.
3. *Enlarge the centers of power and authority*—people's control at the local level and the opportunities for

leadership this creates would lead to the diffusion of power and authority which in Third World countries is vested in a few hands, namely that of the landlords and the bureaucracy.

4. *Lead to the creation of an infrastructure* such as bridges, roads, wells, etc., by the people and those at the helm of affairs, namely the people themselves, would be more interested in providing such facilities.
5. *Strengthen municipal services*—people's councils and committees get involved in local government which will lead to a more stable tax base and enlargement of municipal services.
6. *Reduce the hold of the bureaucracy*—as the people get more control, the role and the power of the bureaucracy, particularly that of the petty government officials, will decline.
7. *Reduce court cases and litigation*—people's participation and control will also lead to the reduction of petty cases in the courts and the lengthy litigations that go with them. As the people's councils become stronger, people can rely more on the arbitration and conciliation efforts of the members instead of taking their feuds to the courts.
8. *Lead to greater cohesiveness*—activities such as those described above would lead to greater cohesiveness within the community and to lesser factions.
9. *Reflect the felt needs of people*—as large numbers of people would be involved in these institutions directly and indirectly, the programmes and activities they will undertake would reflect the felt needs of the people. If agriculture is their main concern, the council's activities will reflect this need. If credit is what the people want, it is possible for such institutions to form a credit cooperative or to establish liaison with the agriculture credit banks.
10. *Enable people's determination of the parameters of development*—if the institutions determine their own priorities, it means that they are determining the kind of development they wish to have and the risks they wish to take and the trade-offs they are willing to

make. We somehow equate the terms development and modernization and wish that societies in the process of becoming modern shun tradition and communal mores, look at issues and problems from an economic standpoint and not from a culturally determined normative perspective. We believe that the decision to shun or retain traditional norms and behavior should be made by people themselves rather than by professional experts, indigenous or foreign.

11. *Lead to greater social justice*—it is visualized that over time such control over determination of priorities and use of resources exercised by the people would mean greater social justice as a much greater number of people would be involved in the decision-making processes. In the words of Freire, "the basic elementary criterion is whether or not the society is a 'being for itself'. If it is not the other criteria indicate modernization rather than development".[10]
12. *Achieve integrated and holistic development*—through decentralized institutions, development could become an integrated, holistic venture rather than a piecemeal fragmented effort of one sector of the economy or the attempt of a particular government department to promote its departmental activities. As problems faced by people do not necessarily fall within one sector or one discipline, solutions suggested through the community and efforts to solve problems may use intersectoral and interdisciplinary strategies.
13. *Develop the rural sector*—the major thrust of economic development in most developing countries should be in the rural areas and through the agriculture sector which employs 60 to 70 per cent of the labour force and provides 40 to 50per eent of the GDP (Gross Domestic Product). The developmental model pursued by many Third World countries of rapid industrialization and urbanization has not been successful and has left the rural areas and the agriculture sector far behind. Economic development can be promoted in the rural areas through decentralized institutions that will reflect the needs of the people. Similarly, agriculture

credit could also be organized through the efforts of people's institutions as community norms, values and control can play a very important part in the repayment of credit. People's institutions can also assist in the matter of collaterals. This will help farmers to borrow from established credit institutions rather than becoming a prey of the loan sharks.

Brief Illustration of Institution Building and its Performance

At the conclusion of this paper, we wish to give examples of institution building that have achieved some of the advantages that we have mentioned in the preceding pages. Two cases will be discussed briefly—the communes of China and the Kibbutzim of Israel.

Communes—China

It is not within the perspective of this paper to go into the historical genesis that prompted the setting up of the communes. However, a brief discussion is in order to give the rationale for setting up the communes.

After the communist take-over in 1949, China attempted to industrialize at a tremendous pace. The "Great Leap Forward" was an effort to industrialize China and introduce the backyard iron-furnace movement. To promote industrialization and other reforms, the strategies of centralization and collectivization were adopted. During the period 1949 to 1957, 700,000 collective farms were established (90% of rural households) and land turned over to common ownership.

However, the entire scheme backfired and China's social and economic development received a severe set-back. By 1958, the 700,000 collective farms were transformed into 250,000 communes. By 1960, an attempt was made to further decentralize by bringing institutions closer to people and involving them in planning, decision-making and implementation of programmes and activities. During the period of 1960 to 1962, within the 250,000 communes, 700,000 production brigades were formed, which were further divided into 5 million production teams. Each team was composed of 20 to 30 households.

We will briefly try to show the connection between the

commune and the system of decentralised institution building which we have been discussing. The commune in China is both an economic and a socio-political entity. It performs for its members a wide range of functions including the provision of minimum guarantees, medical facilities and educational programmes. Indeed, in the field of education, the commune not only guarantees each child a minimum level of education but even makes decisions regarding the level and type of education each child will receive. In the sphere of political development, the communes serve as an indirect training ground to share in the decision-making process, to learn tolerance, to listen, discuss, compromise and so on. In other words, they learn democracy as a way of life. The communes also actively engage in political development and training activities through study groups that use Mao's teaching as their political ideology and the communes as an arena to put these into practice.

We would like now to describe some of the achievements of the commune and to show that as models of institution building they have been able to provide a better quality of life to the people as well as to contribute to the economic and social development of China.

Galbraith states that "the Hsu Hang People's Commune, (in Chia Ting County), has 4,694 households, 20,500 inhabitants and 1,688 hectares (roughly 4,200 acres) of land all irrigated . . . less than an irrigated acre per family". This commune has a grain surplus. Small factories turn out about Y 5 to 4 million (around $2.7 million) worth of goods annually, and produce half the total revenue of the commune. He states further that this commune by Chinese standards is highly mechanized.[11] Myrdal and Kessle state :

> The average grain consumption was 215 kg per person per year. Of this 172 consisted of a basic guarantee, paid out to everyone old as well as young, whether working or not. This allocation was a citizen's right, and did not stand in direct relation to how much work he or she did. Income from work was additional to this allocation. One could say the social aid had been abolished by making basic security the right of every citizen. But this reform had not been carried through without a great deal of

> political work. It was through political work that people's way of thinking had so changed that no one was even thinking in terms of concepts such as 'aid'.[12]

Galston measures the progress of the Lugou Ch'iao (part of Marco Polo Bridge People's Commune near Peking). He comes to the conclusion that between 1958-1972, the amount of vegetable produced rose from 1,000,000 catties to 2,400,000 catties; grain from 700,000,000 to 850,000,000; pigs from 9000 to 481,000; average salary from 300 yuan per year to 400 to 600 yuan; medical facilities from almost none to one hospital, 20 clinics and over 450 medical and paramedical personnel; schools from 10 primary and no secondary to 19 primary and 5 secondary; factories from 0 to 6; fruit from 0 to 680,000 catties; and tractors and trucks from 0 to 30, riding and 68 walking style tractors and 68 trucks.[13]

An analysis of the literature on the work of the communes eloquently testifies to the fact that throughout the process of associating the people at every level—the commune, the production brigades and production teams—i.e., through decentralized institution building, the people are assured a better quality of life as measured by their input in the decision making processes, the medical and educational facilities, the security and sense of belonging that one derives through community membership, participation and involvement in community activities. In this process, besides the emphasis on agriculture, decentralized industrialization has also taken place (a) without the movement of people to urban areas; (b) at the pace that people want and (c) by dealing in products that fulfill the immediate needs of people.

Through collective activities which bring in members of other communes to deal with a problem, the Chinese people have been able to increase the productivity of their land, labor and other resources, face natural disasters—drought, floods, famine, eroding of dykes, etc.—without going under as they now plan ahead for dealing with such disasters. In the Huadong commune (North Canton) a hydroelectric station built as a cooperative project by two neighbouring communes is part of a huge network of irrigation and flood control. Three large dams, each with its own small hydroelectric station, serve the area.[14]

Even writers not totally appreciative of China conclude "the goal of new China is not simply to overtake the West. Mao has set a new direction for his people; its goal is a new man, a new society of selfless equals, which may turn out to be a model not only for the third world but also for the rest of mankind".[15]

Our contention is that one of the most important factors in the progress of China is its strategy of decentralized institution building which has given people the opportunities for participation, involvement and control (PIC) in matters that affect them even within a system which is highly centralized and controlled by the Party and some selected people at higher echelons.

Kibbutzim—Israel

The other illustration that we wish to give is that of Kibbutzim of Israel. The first Israeli communal settlement was established in 1910. By 1966, the Kibbutz movement had grown to 225 communal settlements with a total population of over 90,000 men, women, and children. The origin of the Kibbutz goes back to the revolt of youth groups, particularly of Eastern Europe, against their conditions in the ghettos (*shetetl*). Besides, the youth groups yearned for a more egalitarian society which would not brook discrimination against women too. The socialist ideology "from each according to his ability; to each according to his need", is the ideology on which the Kibbutz movement was established. It aimed at a classless society where means of production would be owned collectively.

The Kibbutzim thus resort to direct participation, involvement and control of people in all affairs that affect them. It achieves this task by organizing all the inhabitants of the commune. All people within a Kibbutz form the General Assembly, which elects a secretariat. The secretariat acts as the executive arm of the General Assembly. The composition of the secretariat varies in different Kibbutzim but comprises generally the secretary, treasurer, farm manager, work organizer and committee chairman. All positions in the Kibbutz at every level are rotated, every two years, hence no person can use his status/position for personal gains. Committees are formed to take care of specific matters such as an education committee to plan and organize education; the health committee for similar work in the health field and so on. Each branch of the economy, for example; animal husbandry, agriculture, warehouse, etc., has a branch manager.

It may be borne in mind that almost 50 per cent of the members of the Kibbutz are involved in one committee or the other at a given time. These committees and other responsibilities are rotated and instead of people competing with each other for these positions they have to be persuaded to assume these positions. People are more conscious of responsibilities rather than of power.

Members of the Kibbutz decide what work they choose to do. By and large, they are given the work that they have chosen to perform. Work that no one likes to do is again discharged by rotation. Young and old, man and woman, children and handicapped are all looked after from cradle to grave if one may use this idiom. Everyone works as best as he/she can and his/her needs are fulfilled within the resources of the community. The members seem to have their physical, financial, emotional, spiritual and other needs fulfilled. They have a sense of belonging and security that is usually the product of living within a cohesive community. It is interesting to note that the Kibbutz and the Moshev (a little different from the Kibbutz) produces 70 per cent of the agriculture output of Israel. The Kibbutz has raised the output per worker four times in 20 years. A 1960 study shows that during the years 1953-1958, Kibbutz labour productivity (gross agricultural output per Kibbutz average worker) increased at an average value of 12.2 per cent annually while Israel's agriculture as a whole rose by 7.3 per cent. Kibbutz dairy herds give yields which are the highest in the world while in 1969 its poultry contributed 21.7 per cent of the eggs with only 16 per cent of laying hens.[16]

Spiro, describing the Kiryat Yeddim Kibbutz, states that "its total gross income is about 250,000 lire. This, it is apparent, is big business. Agriculture is completely mechanized in Kiryat Yeddim, which owns ten tractors, three combines and three trucks."[17] "The Kibbutz knows no slums, economic insecurity and disadvantaged families of low socio-economic status".[18]

The Kibbutzim are a good example of how people's participation, involvement and control exercised through their own decentralized institutions provide not only a better quality of life to the people but also support and enhance the overall Israeli economy. The successes of the Kibbutz seem greater when one realizes that it is based on the principles of equality and cooperation in a country where the socio-political climate is unequal and competitive.

It is not possible in this paper to discuss the case studies fully. A comprehensive analysis will bring out more clearly how the process of people's direct involvement has enhanced the quality of life and how political development is attained through responsible citizen action. Participative democracy through the strategy of institution building can provide a relatively more conducive environment for the socio-political and economic development of Third World countries with less inequality and greater social justice.

REFERENCES

1. V.M. Sirsikar, *Political Behaviour in India : A Case Study of 1962 General Elections*, Manaktalas, Bombay, 1965, p. 253.
2. Fred. R. Mehden, Von Der, *Politics of the Developing Nations*, Prentice Hall, Inc., 1964, p. 118.
3. W.W. Rostow, *Politics and Stages of Growth*, Cambridge University Press 1971, London, N.W. 1, p. 268.
4. H. Halsey, "The Education of Leaders and Political Development in New Nations," from L. Richard Merrit, and Stein Rokkan (ed.) *Comparing Nations*. Yale University Press, 1965, p. 207. The reference to Macaulay is from J. and R.H. Useen, *The Western Educated Man in India*. Dryden Press, New York, 1955, p. 14.
5. Rupert Emerson, *Representative Government in Southeast Asia*. Harvard University Press, 1958, p. 11.
6. Bernard E. Brown, *New Direction in Comparative Politics*. Agra Publishing House, 1962, p. 25.
7. Mehden, op. cit., p. 22.
8. Sharan Parmatma, "Thoughts on Democracy" from *Political Studies* by Tiwari and Sharma, Shivalal Agarwala and Co., Agra, 1966, p. 187.

9. Barbara Ward, *Nationalism and Ideology*, W.W. Norton and Co., Inc., New York, 1966, p. 106.
10. Paulo Freire, *Pedagogy of the Oppressed*, Continuum Books, Seabury Press, New York, 1971 p. 161.
11. John Kenneth Galbraith, *A China Passage*. Houghton Mifflin Co., 1973, p. 106, 107.
12. Jan Myrdal, and Gun China Dessle, *The Revolution Continued*, Pantheon Books, 1970, p. 53.
13. Arthur W. Galston, and S. Savage, *Daily Life in People's China*. Thomas Y. Crowell Co., New York, 1973, p. 105, 106.
14. Committee of Concerned Asian Scholars, *China : Inside the People's Republic*, Bantam Books, Inc., 1972, p. 154.
15. Klaus Mehnert, *China Returns*, E.P. Dutton and Co., Inc., New York, 1972, p. 236.
16. Dan Leon, *The Kibbutz : A New Way of Life*. Pergamon Press, Oxford, 1969.
17. Melford, E. Spiro, *Kibbutz: Venture in Utopia*. Schocken Books, New York, 1963.
18. Mordecai Kaffman, "Characteristics of the Emotional Pathology of the Kibbutz Child," *American Journal of Orthopsychiatry*, Vol. 42, No. 4, July, 1972, p. 710-716.

SIX

ALTERNATIVE RURAL STRATEGIES
Modus Operandi and Constraints, Representation and Participation in Multi-level Decision-Making

Kurt Dopfer

RURAL SYSTEM DEFINED

THE rural world—its specific way of life, modes of production and consumption, institutions and behavioral rules—may in its complexity be best conceived as a *system*. A system can be said to be a set of elements standing in relation to each other, say in interaction or interdependence. The elements of the rural system are a set of socio-economic institutions, patterns of resource distribution, modes of production, and behavioral rules with their related norms of sanctions and rewards. The *rural system* may, thus, be defined as a set of the above stated elements that stand in interaction or interdependence.

RURAL SYSTEM IN OPERATION

The definition put forward leads to the first principal question : how do the various elements or variables of the rural system interact? An answer to that question requires that we broadly distinguish between (1) the causal structure that shows how the variables are related to each other, and (2) the hypotheses that spell out the "direction" of causality among those variables.

The following discussion lays out the basic causal structure and the causal direction of the variables of the rural system. The causal structure of a rural system may be reduced to two major variable sets. They are respectively (A) the mode of production and the prevailing production techniques, and (B) the institutions. This includes the type of appropriation of the means of production (particularly land), the ideologies in power and modes of behaviour. A certain resemblance to the Marxian classification of the societal factors is apparent. However, instead of assuming that the variable set B is determined by variable set A, we think that the hypothesis of a reciprocal causal interaction between both ontological spheres A and B is a more fruitful starting point for the analysis of the rural systems prevailing in contemporary Asian countries.

The two major sets of elements and the reciprocal relationships that constitute the rural system are depicted in Fig. 6.1 The two arrows, one reaching from A to B, the other from B to A, indicate the reciprocal interaction among the two element sets.

FIG. 6.1

Rural System without Strategy Linkages

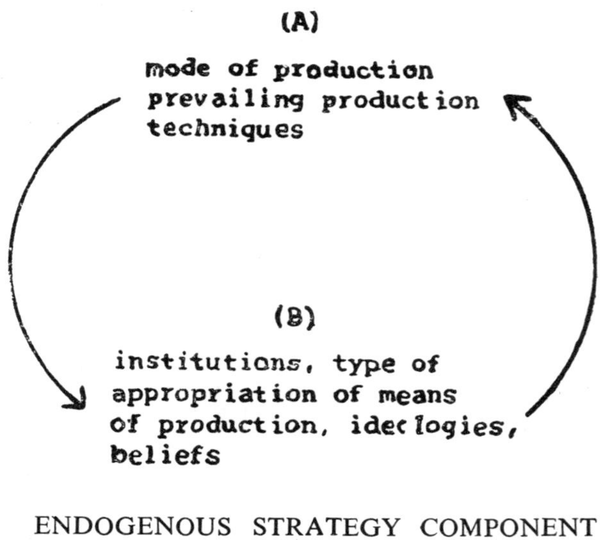

ENDOGENOUS STRATEGY COMPONENT

The system above is typically an endogenously determined rural system. In this context, strategy does not come into the picture explicitly—at least, if strategy is conceived as an "exogenous" push, e.g. an intention to introduce into the system new ideas, motivational stimuli, organizational capacities, resources and so forth. However, it is conceivable and probable that a strategy is actually being pursued, even though the objectives and forms of implementation may not be spelled out explicitly. Such a strategy would constitute an "endogenous" strategy as contrasted to an "exogenous" strategy (with its strategy linked to an outside decision-making body). A strategy in an endogenously determined rural system will usually be pursued by those people who hold the socio-economic power in the community. The strategy will be designed in a way that the status quo of the power structure, its resource distribution and ideological control remain preserved. It, thus, reinforces the existing relationships between A and B. Many historical instances reveal that a small initial push, say a change in the resource conditions that affects the modes of production, suffices to bring about a cumulative process of reinforcing interactions that result in a highly unequal resource distribution, which, in turn, generates the political power that maintains that regime.

In Fig. 6.2 the two components constituting the *strategy context—representation* and *participation*—are introduced to the endogenously determined rural system. The latter is denoted by stratum X, the strategy context by stratum Y. The two elements of the endogenously determined rural system, the mode of production and the institutions determine the strategy context. viz. representation and participation. The relationships are depicted by two pairs of arrows ranging from the two elements of the socio-economic system to the strategy context. Thus the endogenously determined strategy context of the rural system determines the content of the strategy actually pursued in the rural system and as such shapes the socio-economic system of which it is an outcome. The pair of arrows reaches from the endogenous component of the rural strategy to the socio-economic elements (A) and (B) of the rural system.

The basic difference between an endogenously determined rural system with and without an endogenous strategy lies in the fact that the former system is likely to be more difficult to change—if change is to bring some "disturbance" into the prevailing system—than the latter. As "disturbance" we may

FIG. 6.2
Rural System with Endogenous Strategy Linkages

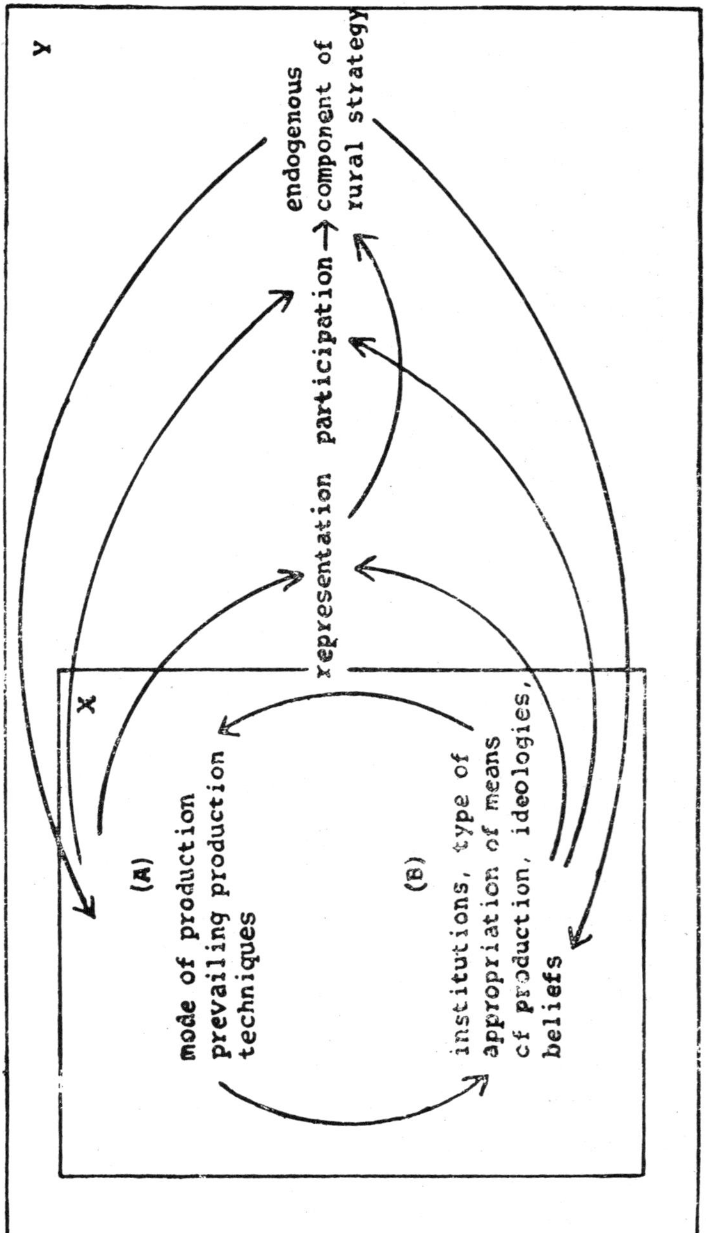

consider any event that jeopardizes a given pattern of socio-economic distribution, the established power structure, and the welfare generated by the prevailing patterns of power. The disturbance will become politically relevant if the social group that is in a position to interpret the event as a disturbance is strong enough to draw and subsequently implement its respective conclusions. It will, for instance, formulate an "alternative" strategy that counteracts any exogenous disturbance that threatens the "stability" of the system.

EXOGENOUS STRATEGY COMPONENT

This brings into focus the second element of a rural strategy—the exogenous element. If a rural strategy appears in a context higher than the local level (higher than that of the endogenously operating rural system), it will constitute an exogenous factor to the rural system. A national or overall rural strategy is, thus, never a homogeneous entity. It is always composed of a set of existing forces that are (1) an inherent part of the rural system (be they reinforced by deliberate endogenous strategies or not) and (2) exogenous forces introducing some form of energy for developmental change. It does not seem appropriate to view these two different elements only as two different levels in decision-making. These are two different levels in terms of deeply vested, and mostly antagonistic, interests. The reapportioning of decision-making powers in a multi-planning context, thus, entails much more than re-evaluations in terms of informational and allocational efficiency in a centralization vs. decentralisation scenario. It requires genuine political evaluation and action.

OVERALL RURAL STRATEGY IN ACTION

Structural Context

In the following, the causal structure linking the endogenously determined rural system to the larger national context of decision-making shall be given. In Fig. 6.3 the two sets of socio-economic variables and the endogenous strategy element constituting the endogenously determined rural system are linked with the central, national, i.e. exogenous strategy element.

Stratum Z encompasses the set of all variables that are exogenous to the rural system as well as by the endogenously determined rural systems denoted by X (rural system without endogenous strategy component) and Y (rural system with endogenous strategy component).

In the larger national context, the strategy component as defined by a specific form of representation and participation receives a dual meaning since it pertains to the local level as well as the national level (or any other level above the local level).

Once the postulate is accepted that *alternative* strategies should be pursued (as is suggested in the current discussion on rural development and rural strategies), the crucial problem is how can *combined* (national *and* local) forms of representation and participation be theoretically ascertained and actually implemented.

Before embarking on a discussion, let us first of all define the terms "representation" and "participation". *Representation* may be defined as the ratio of people actually engaged in the pursuit of a specific activity to the number of people required to pursue that activity adequately ; for instance, the type and extent of actual representation of a social group *vis-a-vis* the representation required to articulate and pursue the interest of that social group. *Participation* is defined as the ratio of actual involvement over involvements required to effectively pursue specific activities, at a given level of representation for the group concerned. The "efficiency" of a specific type of strategy context, i.e. the "political system" or the "organizational matrix in which decisions are made and interests are pursued, can adequately be evaluated only with a composite index that includes both the values for representation and participation. Good representation may go with bad participation, and *vice-versa*. It seems that the two components have been mixed up in the recent discussion on (rural) development. Either the two were assumed to be identical or it was tacitly conjectured that if only participation (representation) can be increased representation (participation) will automatically result. Though there is, as will be shown, a causal inter-dependence between the two, there is no necessary link between them.

Given the two levels of decision-making (local, national)

FIG. 6.3
Rural System with Endogenous and Exogenous Strategy Linkages

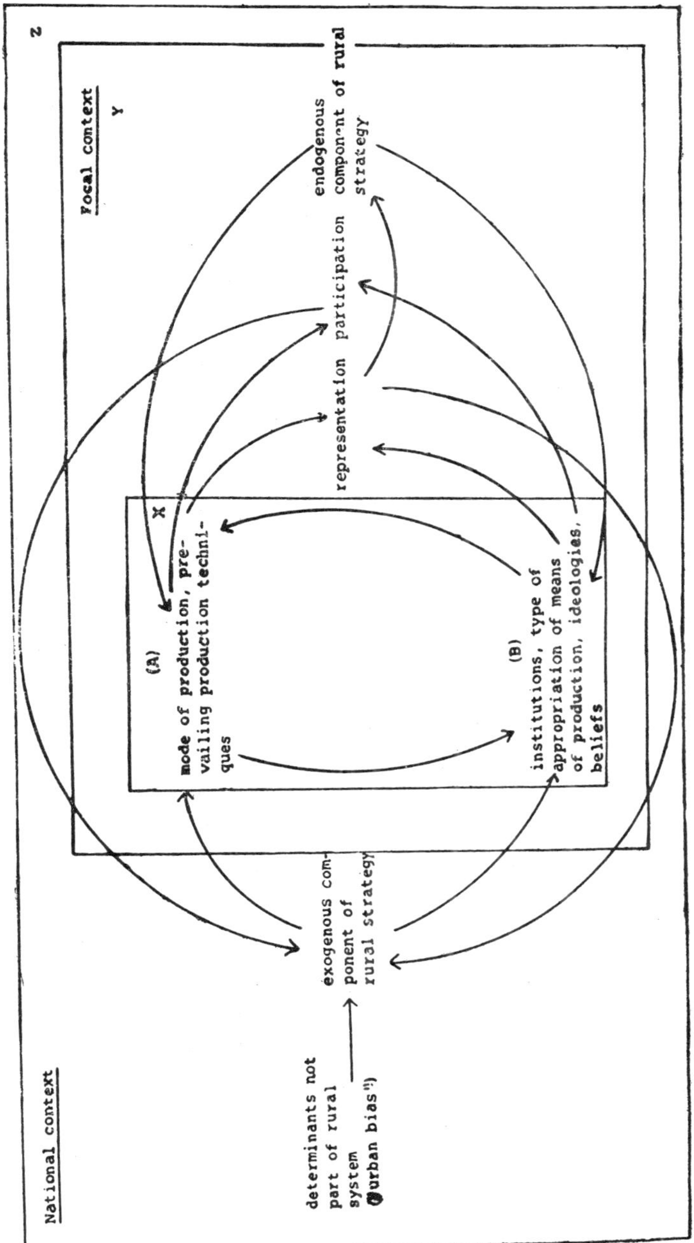

and the two components of a strategy context (representation, participation), a simple classification results. We have respectively :

(1) representation at the national level,
(2) representation at the local level,
(3) participation at the national level,
(4) participation at the local level.

Content of Rural Strategies

The Socio-economic Strategy Component

The preceding discussion suggests that a rural strategy is not confined to the socio-economic strategy proper but includes a political component as well. Thus, the content of a strategy, at least if the present state of affairs in the rural areas of Asian countries is discussed, is necessarily composed of two elements: first, one that pertains to the socio-economic strategy in its narrower sense, and, second, one that pertains to the political context in which the strategy is bound. Both components of a rural strategy may be *status quo* supporting, or, they may favour an *alternative* state of the system.

In discussing the first component of a rural strategy, a list of the major objectives characteristic of that component is given. The list is confined to alternative rural strategies (omitting, thus, the *status quo* supporting strategies) ; as such it is a summary of the objectives as spelled out in the ongoing discussion on rural development and strategies. Some of the major objectives are :

- redistribution of economic resources
  - land redistribution, land reform
  - more egalitarian access to inputs, such as HYV-inputs, i.e. fertilizers, energy
  - access to financing capital
  - access to marketing facilities
  - access to education
  - access to other socially provided services
- employment creation
  - direct attack on unemployment

    - employment programmes
    - appropriate technology
    - labour intensive production techniques
- redistribution of welfare
    - redistribution of income and of provisions in kind
    - minimum income level
    - provision of basic needs
- restructuring production
    - change in output pattern
    - wage goods production
    - ecologically adjusted production (and consumption)
    - intertemporal re-evaluations of non-renewable resources
- reassessment of the entire production process
    - self-reliance
    - de-alienation
    - increased representation and participation (in socio-economic and political realms).

Three comments may be helpful for the assessment of the objectives :

First, most objectives stated are closely interrelated with each other. For instance, income distribution can only be achieved by a more egalitarian distribution of the resources (not by secondary income distribution such as taxation and income transfers). Similarly an increased application of labour intensive production techniques will not only contribute to the solution of the employment problem but also to the production of wage goods which constitute the commodity basket appropriate to meet minimum or basic needs.

Second, the objective of self-reliance is likely to be attained if the above set of objectives (not referring to self-reliance) can be attained. It does not seem appropriate to start off with the over-arching postulate of increased self-reliance, and *a priori* assume that the attainment of the rest of the objectives will follow. This relationship is not well enough established to warrant such a hypothesis. As stated in comment one, it seems of primary importance to theoretically understand the close causal interdependence of the objectives and processes resulting from the actions appropriate to attain them. In a context of causal interdependence, priorities among normative and instrumental variables lose much of their meaning.

Third, increased representation and participation have in themselves *normative* significance. They are not only intrumental in terms of contributing to the attainment of the set of other objectives. The nation of "efficiency" of various types of representation and participation (as used earlier) becomes somewhat vague. A course of action emerging from a specific type of representation and participation may be less efficient in terms of attaining a specific set of objectives than alternative courses of action, but a high normative significance may be attached to the prevailing (e.g. egalitarian) type of representation and participation compensating for that loss in efficiency.

Improved Representation of Rural Population at the National Level

An even more intricate and theoretically even less understood problem is this : How can a type and level of representation and participation be secured that allows for the formulation and implementation of alternative strategies ? Following from the classification presented in paragraph 1, our first problem is the lack of appropriate representation of the rural population at the national level of decision-making and planning. A first clue to the phenomenon of under-representation may be the prevalence of unsatisfactory organisational forms of representation, e.g. types of political systems in which an uneven distribution of representation is *system-immanent.* That is to say, the political system is undemocratic (however this may be defined and interpreted in a specific historical context), and hence no adequate representation of the majority of the population may be expected. While there are various examples for such systems in Asian countries, establishing more democratic forms will not lead *per se* to improved representation Besides the fact that the reorganisation of the political system in a direction of improved representation for the yet under-represented is itself a highly political act, and as such calls forth the very forces that constitute the undemocratic system. Indeed, there seem to be system-inherent forces at work which impede the actual performance of political systems directed towards the needs of the rural population—whatever range of democratic representation of the rural population may prevail.

An explanation of this phenomenon is conceivable if the rural system and the rural strategies (both the endogenous and exogenous components) are viewed in the large context of the overall nation or society encompassing both the *rural* and *urban* areas. Decision-making is usually done in the cities. Government and governmental institutions are located in the cities and bureaucrats live in the cities. Irrespective of problems of informational inefficiency due to the distant link between rural and urban areas in terms of space and time, there will be—almost by necessity—an "urban bias" in any strategy pursued considering the different interests of the urban people. The question then is what type of rural strategy will be favoured by the urban elites. Moreover, which social groups in the rural areas will serve better the cause of the urban interests? Are alternative strategies that lead to improved basic needs then conducive to urban interests, viz. of obtaining a sufficient supply of agricultural commodities that fit the urban standards and tastes? Or, are the prevailing strategies and modes of operation of the existing rural systems more conducive to the needs of the urban population? More specifically, will existing land owners and related rural elites serve the interests of the urban people better, or, will a rural population that is better represented in both the socio-economic and political realms defend the cause of urban interests?

Indeed, such considerations lead to the very core of an alternative rural strategy. How can any change in the national system of representation be expected or be advanced if not from forces that are typically exogenous to the rural system; and, if these forces have to protect their own interests (which are not necessarily those articulated in the system they want to bring about), how can these forces be expected to serve as a basis for alternative strategies that would serve the cause of the large rural population? It is at this point that the quest for decentralised decision-making gains momentum. But, it should be recognised that the change of the total system of representation itself is a highly political act that is unlikely to come about immediately or, in some cases, easily.

Improved Representation of Rural Population at the Local Level

Assuming that decision-making powers are increasingly delegated to the local levels, the question arises how to establish and implement a system that represents those social groups that benefit from an alternative strategy at the local level. Various proposals have been made in recent discussions. They include increased representation in the decision-making process in the socio-economic realm through the formation of cooperatives for (1) securing the supply of inputs, e. g. for HY-varieties (producer cooperatives), (2) for adequate supply of financing capital (rural cooperative banks), (3) for improved marketing to sell the tradeable surplus and to ensure an increased supply of non-agricultural goods (marketing cooperatives).

A useful distinction may be introduced by classifying the cooperatives into those that are divisible and usually individually produced and consumed, and those that are indivisible and usually socially provided and collectively consumed. The latter encompasses various types of what are traditionally understood to belong to the infrastructure and governmental or community sectors such as, education, transportation, etc.

In some cases, a reapportioning on the relative importance of these two areas, may prove helpful in meeting the basic needs standards. In others, it may not. This issue requires further investigation. The euphoria in the present results from expectations resulting from a shift from the first area (divisible and individual entities) to the second (socially provided and consumed) seems premature.

Irrespective of the relative importance of the two areas, the problem of adequate decision-making, i. e. the formulation of objectives and adequate courses of action, remains to be solved. One proposal for improved decision-making at the local level may be the use of *direct* forms of decision-making as practised for centuries, in provinces, cities and villages of Switzerland and more recently in the Peoples' Republic of China. The latter received much attention in recent discussions; I shall therefore briefly focus on the former.

We may define *direct democracy* as any decision-making process in which each member of a society has a vote not just for electing representatives who decide issues on their behalf but also to vote on the issues themselves. They articulate their preferences on the various issues *directly*, and not indirectly through their representatives. If, for instance, in a Swiss village a new kindergarten or school is being considered, the population is asked to vote on the issue. If the votes show an absolute majority (i. e. over 50% are in favour), the kindergarten or the school will be built.

A first step towards improved decision-making at the local level may consist in the mapping out of those areas of action in which such a procedure of decision-making may be applied. The two areas as outlined above may, indeed, merge sometimes in terms of similar *modi* of decision-making. In some instances, combined forms, encompassing individual and collective decision-making, are conceivable. For instance, for the distribution of available inputs (including land) both the capacity derived from individual purchasing power and the preferences that are spelled out in a process of direct collective voting may be taken into consideration, (say, once in a year or once in a harvest). In any case, the norms ruling such social distribution could be laid down by the local community.

The proper domain, however, is the second area of societal activities and outcomes. The establishing of new forms of direct collective decision-making at the local level in this area is likely to show immediate satisfactory results, and will, as such, have an inducement effect even in areas for which it was not originally designed.

Thus, my proposal would be to start a rural strategy with improved forms of direct representation in the area of those activities that are by their very nature conceived by the community as a whole. Subsequently, we can extend the procedure to other areas in which decision-making capacity is also required (for the attainment of outcomes stated in the set of alternative objectives) but is yet lacking. In other words, the building of new institutions should be conceived within a sequence of strategic steps that allow the effects from "learning by doing" to become as effective as possible. The ideas of "inducement mechanisms" and of "learning by doing" bring us to the second major aspect viz. participation.

Improved Participation of Rural Population at the National and Local Levels

Given the distinction of national and local levels, participation may be classified into:

(1) participation at the national level, and
(2) participation at the local level.

The problem of participation at the national level is intrinsically linked with that of under-representation at the national level. Effective participation of the rural population is limited and the existing possibilities of participation are further decreased because felt under-representation frustrates participation. Once the causal nexus between the possibilities offered by representation and participation becomes likely, any adequate change in the system of representation will generate a high level of participation. This hypothesis may in some empirical instances fail to materialize or be verified. But in other instances, the inducement mechanism implied in that causal nexus may well bring forth additional involvement in political decision-making from groups thus far believed to be uninterested, fatalistic and "traditonal".

The proposition to decentralise implies not only that institutions of decision-making (e. g. direct democracy) should be increasingly located at the local level but also that the participation be increased at that level. This pertains to the political-context component as well as to the socio-economic component of a rural strategy. As for the former, the first major act intended to increase representation and participation of the rural population at the national level may well be the last: the delegation of decision-making powers to the local levels. More political decisions may have to be made in the long run perspective of development. Contemporary societies do not usually remain rural and agrarian forever. They grow, at different paces, into societies with different outlooks, through the provision of more non-agricultural commodities, increased freedom from toil, and increased leisure. Higher levels of environmental awareness (or, at least, of education) develop, though a few of our Indian colleagues, in the agony of past failures, may promote Gandhianism, the "return to the villages", as an optimum *regime* of societal development. However, for the moment, there seems to be no more pressing decision on the national level than the one to implement rural strategies that acknowledge the irrelevance of centralized decision-making and to implement new institutions that increase local decision-making capacities. It is in this dynamic context that the discussion on "local level planning" as spelled out on the agenda of this seminar gains its importance.

Given a scheme of local representation, the *conditio sine qua non* for increased rural participation is given. However, the problem of increased participation is yet far from being resolved. As stated earlier, improved representation is likely to generate what may be called "participatory energies," but perhaps not enough and perhaps not the optimum that could be reached under the conditions of a strategy that stimulates participation.

If reliance is not merely on "soul massage", Sunday speeches, etc. inducement mechanisms must be contemplated that increase the motivational level of the rural population involved in the decision-making process. One major inducement mechanism is given by the close causal nexus between formulation and implementation of a decision. If a decision is made by an individual, that individual will usually be motivated to implement the decision. What is valid for a single agent is, *ceteris paribus* valid for the various decision-making levels, e.g. the local *vis-a-vis* the subordinated levels in the national decision making hierarchy. There will be, moreover, a very small informational loss since, with the locus of formulation and implementation on the same level, the message is conveyed in the shortest informational distance possible with no interlevel frictions or errors. Any success in terms of desired outcomes that results from improved decision-making will further increase the motivational level that, in turn, brings forth further decisions needed to attain the desiderate.

If the dynamics and the efficiency of a thus conceived participation process is considered, two aspects deserve particular attention :

(1) the initial impulse that brings into motion a decision-making process or its acceleration, and
(2) the appropriateness of the informational content, i.e. the quality of the message.

As for the first aspect, the central problem lies, *ex hypothesi*, with the formulation, not the implementation of decisions. If a short or zero distance between the two *loci* of formulation and implementation and a "normal" behavioural response (as stated above) are assumed, an inducement in the formulation of decisions will solve the problem of implementation. Moreover, considering the reinforcement between the two components of formulation and implementation, increased decision formulation will not only bring forth an adequate implementation but the latter will in turn increase the level of the former.

Given this cumulatively reinforcing cycle, the key problem is how to provide for initial "impulses" in terms of decision formulation, e.g. to reformulate the objectives, to conceive alternative courses of action, and so forth. Reference must be made here to the earlier discussion on *direct collective* decision-making. If the decision-making capacity of the rural system is to be increased, an exogenously initiated disturbance of the system in the form of a collective decision-making push may be an appropriate strategy. It might not appear most efficient in terms of costs incurred by the informational distance (as defined before), since the individual (in contrast to the collective) decision-maker has an informational distance with the value of zero. However, in many cases, collective decision-making procedures will bring about decisions that otherwise would not be made at all. It will result for the first time in decision-making processes, and as such induce awareness. It will stimulate reflections on alternatives and the appropriate courses of actions to be taken.

Given the motivational pushes to decision-making, the second aspect comes in as decisive momentum : Is the content of the decision appropriate ? Even if the level of decisions is high and the informational efficiency is perfect (i.e. no loss in conveying the information, the decision formulation to implementation), the decision itself may be inadequate for the attainment of a desired outcome. The issue of appropriateness of the content of decisions is related to the appropriateness of alternative development strategies as such. The discussion, however, may also be given a more specific and, at least in the context of our discussion, a more operational meaning. Micro decisions must prove their rationality in the macro context of which they are part. If the individual producer household of a rural community is the micro unit, the macro context will be the local level. If the rural communities are taken as the micro units, the macro unit will be the national economy. Except in the rare cases of a simple village economy where no exchange transactions exist (i.e. not even at the village level), the macro consistency of socio-economic decisions, activities and outcomes constitutes the performance indicator *par excellence* of the total system.

The basic objection raised against the participatory approach is that the micro decisions that emerge from grass roots decision-making are unlikely to be well coordinated at a macro level, i.e. at the national, province or district level, and that, consequently, an overall coordination should precede any activity at the micro levels, i.e. the levels of the village, the producer households and the individual agents. To bring this issue into proper perspective, it seems necessary to clearly distinguish between two major aspects of the socio-economic process and planning. The first aspect is the resource relationships (stocks, flows); the second, the set of activities (as determined by social norms, motivations, exogenous stimuli, etc.) that correspond with or underlie the resource relationships. A closer account of the objections aired against the proposed participatory approach reveals that they rely on two gross assumptions :

(1) the overall transparency of resource relationships can be obtained and explicitly stated in a national plan, and
(2) the resource plan will bring forth a set of activities necessary to implement the resource plan.

Neither of the two gross assumptions—on theoretical or empirical grounds—is warranted. As for the former, it has become increasingly clear in the last two decades that informational efficiency (information gathering, informational exchange among various decision making levels, and decision delegation) is difficult and often impossible to achieve in developing countries. The state of "under-development"—defined by the discrepancy between the potential and actual state of a society—is a general expression for the lack of just those institutions and performance qualities on which the traditional planning approach heavily relies. Advanced coordination techniques such as national balance sheets, integrated sectoral planning, input-output tables and econometric (macro and linear programming) models seem to have supported the conclusion that the paper plans thus derived must necessarily find their counterparts in the real world. Empirical evidence has established that they do not.

The second assumption seems equally unwarranted. Past experience has demonstrated that the announcement of a national resource plan and its delegation to the lower strata of the decision-making hierarchy do not provide the *impetus* to actually generate the required type and level of activities. This was apparent in countries where reliance was on quite sophisticated resource plans such as the Indian plans. But it was also apparent in the Eastern European countries (with their advanced informational and organizational structures) where even a system of centrally monitored incentives could not induce appropriate types and levels of activities. Finally, there was reliance on forms of genuine decentralized decision-making. The experience with approaches that start off with overall consistency postulates and assume the levels of activities to be (positively) correlated with the degree of technical sophistication of the resource plans, hardly warrants the neglect of the potential implied in the outlined representational-participatory approach. The rationale for this approach lies, first, in the high informational efficiency due to the small loss in information transmittal (unity of decision formulator and implementor), and, second, in the motivations that are likely to be generated once the appropriate strategies for improved representation and participation are applied.

Thus, the priorities between the two basic postulates of a development strategy should be redefined and reversed. Priority should not be given to overall resource consistency supplemented with the belief that the activities that constitute that consistency will come forth. Instead, the sequential logic should rely on priority given to the decision-making potentials as generated at local levels by alternative representational-participatory approaches. Within such approaches, coordination would have the nature of an *anticipated response* to actually emerging activities.

CONCLUSION

It goes without saying that many problems alluded to in this paper will require substantial additional efforts to be dealt with more appropriately. Yet unsatisfactory as the present state of the art may be, there seems to be an ever growing *consensus* within our academic community toward the direction taken in this paper. As this consensus grows, policy making may also move in that direction—assuming that Keynes was right in his assumption that the world is "ruled by ideas, and by little else".

The divergent views in this paper pertain to specific issues and to some clarifications. Their summary may serve as an *agenda* of unresolved issues or of issues that yet require more analytical (theoretical and empirical) depth. Put into the form of postulates they are :

— *First*, a clear analytical distinction should be made between representation and participation. The respective confusion in the current discussion may prove counter-productive.

— *Second*, the decision-making processes on the local levels should be more deeply contemplated. The simple approach of undiscriminatingly taking decisions out of the hands of individual decision makers and delegating them to poorly defined and understood collective entities may add another page in the book of development failures.

— *Third*, the rural system should be conceived, defined and analysed in its dynamic properties. Static visions of a rural community may serve least the interests of the villagers who may have little understanding for static concepts of idealised village life, but have a vision beyond the mere necessities that in the long run are unlikely to be satisfied within the confines of a rural system. The high rates of rural-urban migration in

Asia (and elsewhere) are but one empirical indicator of this hypothesis.

— *Fourth*, emphasis should be placed on the initiation and sustenance of activities that are carried out at the (local) levels of the rural system on a large scale basis. Alternative rural strategies should be designed to provide a new *modus operandi* in terms of new institutional settings and forms of representation and participation. Given this critical exogenous push to the rural system, activities will emerge that must, in turn, be anticipated on a higher, say national level, and coordinated in accordance with national priorities and the rationale of overall allocational efficiency.

A final note : The near future may provide some indication that Asian development is not destined to be as dramatic as the present awareness of past failures suggests. The 1970s, seemingly intellectually poor and economically frustrating years, may in *retrospect* prove to be better than its reputation. The alternative development strategies currently discussed may provide a dynamic *momentum* that we, as economists or social scientists with our yet limited theoretical concepts and tools, are neither able to adequately predict nor explain. This hope, after all, justifies our search for alternatives.

SEVEN

REPORT OF THE CONSULTATIVE COMMITTEE ON ALTERNATIVE STRATEGIES FOR DEVELOPMENT WITH FOCUS ON LOCAL LEVEL PLANNING AND DEVELOPMENT

INTRODUCTION

Organization of the Meeting

The Consulative Meeting on Alternative Strategy for Development with Focus on Local-level Planning and Development was convened by the United Nations Asian and Pacific Development Institute (UNAPDI) in collaboration with the United Nations Economic and Social Commission for Asia and the Pacific (ESCAP), and the Interagency Task Force on Integrated Rural Development. It was held at Bangkok from 31 October to 4 November 1978.

Attendance

The meeting was attended by eight experts in their individual capacity, from the region.

Representatives of the following United Nations bodies were present : Economic and Social Commission for Asia and the Pacific (ESCAP), United Nations Headquarters, United Nations Children's Fund (UNICEF), and United Nations Development Programme (UNDP).

Representatives of the following specialized agencies attended : International Labour Organization (ILO), Food and Agriculture Organization of the United Nations (FAO), United Nations Educational, Scientific and Cultural Organization (UNESCO), World Health Organization (WHO).

The Meeting was also attended by the Director and staff members of ILO/ARTEP, UNAPDI and representatives of Agricultural Development Council (ADC), and Asian Institute of Technology (AIT) and two experts from outside the region.

The list of all the participants is given in Annex.

Election of the Chairman

Mr A.Z.M. Obaidullah Khan was unanimously elected Chairman of the meeting.

Adoption of the Agenda

The meeting adopted the following agenda :

1. Election of the Chairman
2. Adoption of the Agenda
3. Statement by the Director of UNAPDI
4. (a) Identification of main elements and issues and review of innovative measures in new development strategies/approaches.
   (b) Clarification of concepts and issues in new development strategies/approaches.
5. (a) Delineation of the role and scope of local-level planning in the context of new approaches to development.
   (b) Issue for coordinating local-level planning with multi-level planning.
6. Suggestions for further work to be done at the regional and country levels.
7. Adoption of the Report.

Brief Account of the Proceedings

Inauguration

The meeting was inaugurated by Mr J.B.P. Maramis, Executive Secretary of ESCAP. In the inaugural statement, Mr Maramis welcomed the participants and informed them that the Consultative Meeting was the first of a series of two meetings on the important subjects of alternative development strategies and local level planning for predominantly rural economies of the Asian and the Pacific region with focus on the poor and their participation in the processes of development. The second meeting sponsored by the Agricultural Division of ESCAP Secretariat would follow immediately after the conclusion of the Consultative Meeting.

The Executive Secretary stated that "eradication of poverty, unemployment and social inequality" has become a task of first priority in the work programme of ESCAP and associated institutions such as the UNAPDI. The Interagency Coordinated Plan of Action for Integrated Rural Development, approved by the Commission at the thirty-third session in 1977, was being satisfactorily implemented under the supervision of the Interagency Committee on Rural Development for the ESCAP region.

The Executive Secretary felt that there was need to re-examine the concepts, strategies and programmes in rural development with reference to the objectives sought namely, self-sustained growth with equity, greater employment and improvement in the quality of life of the poor with their full participation as equal partners in development. He expressed the hope that the Committee would undertake this task and that it would also formulate practical guidelines that can help governments and development workers in the region in designing viable alternative strategies for development within the socio-economic and political context of their respective societies.

Opening Statement by the Director

In his opening statement, Dr Vinyu Vichit-Vadakan, the Director of the UNAPDI, thanked the participants for having accepted his invitation to participate in the Consultative Meeting. He also informed the meeting that some participants had to cancel their participation owing to heavy duties. He referred to UNAPDI's aide memoire of 10 July 1978 sent with letters of invitation to all the participants. As stated therein, the objectives of the Consultative Meeting were twofold, namely, first, to systematically identify essential elements and clarify concepts and issues in outlining alternative or new strategies or approaches to development, and second, to delineate the role and scope of local or community level planning in the context of new approaches to development.

The twofold task before the meeting was expected to be undertaken on the basis of an analytical review of recent country experiences, and an evaluation of innovative measures and new approaches adopted by the countries in the region, within the framework of the commonly accepted goals of target group-oriented participatory development, both at the micro and macro levels.

The Director expressed the hope that the focus of the deliberations in the meeting would be on operational aspects of designing viable alternative strategies for the developing countries of the Asian and the Pacific region. He also expected that the discussion in the meeting would provide useful guidelines to the Institute in designing necessary training and research activities and in rendering advisory services to interested governments, in collaboration with national and regional expertise and institutions, in designing appropriate country-specific strategies for integrated development.

Daily Programme

On the first day of the meeting after introduction of agenda item 4 by Mr Ram C. Malhotra, Deputy Director, UNAPDI the background papers and country statements were presented in the morning and afternoon sessions. Later in the afternoon session the Chairman invited all other participants to also make statements on main elements and issues and a review of innovative measures in new development strategies.

On the second day, the elements and issues identified in the various statements by all the participants were discussed. Later in the day, the participants were divided in three groups, each group being requested to further explore in-depth the issues and implications involved in each of the following topics :

(i) Basic elements of an alternative approach to development;
(ii) Local Level Planning;
(iii) Linkages and implications of local level planning.

The reports of the three working groups were presented and discussed in the morning session on the fourth day. Later agenda item 6 was introduced and briefly discussed. The group was also informed about UNAPDI's proposed "research-cum-training project on appropriate national strategies for rural development for alleviation of mass poverty, unemployment and social inequality in the ESCAP region" to be undertaken in 1979-80 subject to UNDP's approval of the proposal for its programme support.

The draft report of the meeting was discussed and adopted in the morning session held on the fifth day. The meeting ended with a closing statement by the Director.

Closing Statement

The Director expressed his own and his colleagues' satisfaction with the smooth and productive proceedings of the meeting as reflected in the contents of the report. Given the existing socio-economic structures as parameters for development to take place in the developing societies in the region, he felt that changes will be brought in those structures by the development process itself, provided it has been conceived with analytical perception of the objective realities and a better understanding of the development process itself, in human terms. He felt that very little seemed to be known about the task of development and its numerous facets, even after nearly three decades of development experience in the developing world. We are still scratching the surface of the basic goal of human development. UNAPDI therefore hoped to undertake, in collaboration with the Interagency Committee and Task Force for Integrated Rural Development in the ESCAP region and with national institutions and expertise available in the region, further action-research along the lines suggested by the group in the report. UNAPDI's role, however, is more of a catalyst in providing assistance to countries in the region in their efforts to design appropriate development strategies. UNAPDI strongly feels that the task of development at the country level would have to be done by its own expertise and national institutions. UNAPDI's inputs are mainly in terms of regional experience, again primarily through TCDC or through expertise and institutions available within the region. Finally the Director thanked the Chairman and the participants for the fruitful discussion and successful conclusion of the meeting.

ELEMENTS OF AN ALTERNATIVE APPROACH TO DEVELOPMENT

The International Context

An alternative approach to development must be considered in the context of a world experience to date which shows that development has not succeeded in achieving its objectives in terms of human development. For the vast majority of people conditions have not improved appreciably and for a significant percentage they have worsened. This process of uneven development (including rapid growth in a context of inequity, poverty and unemployment) takes place at both a global level and at national levels, between and within countries. For most countries of the Third World generally, and in the Asian and Pacific region specifically, incorporation into the international economic system has resulted in most cases in the development of internally fragmented economies dependent on conditions in the international economy. These economies usually are characterized by production and consumption structures which derive their dynamism from their specific dependent relations with the international economy. This has found expression in such phenomena as dependence on export of primary products which are competitive on international markets, a frequently high degree of external control of national development processes, increasing indebtedness and increasing dependence on industrialized countries for basic consumption goods.

The incorporation of Asian and Pacific countries into the international economy has often helped create and maintain national social and political structures whose interests are perceived to be centred on the maintenance of the existing system. At the same time, the inherent instability of this system and its inability to create the conditions for lasting development even for the few has impelled the search for alternative development approaches and strategies.

National development strategies must be seen in the context of this international system, the changes in which are the subject of the New International Economic Order and the current North-South dialogue. Clearly achievement of the New Economic Order will be a major objective of the new international development strategy, and the degree to which this is achieved internationally will bear on the success of any national development effort.

The critical importance of this issue was noted even though the meeting was focused on the national level.

The National Context

Alternative national strategies should be seen in the context of a process of socio-economic change that involves the transformation of agrarian society to reach a common set of development goals focused on the capacities and needs of people. These goals include a nationally-determined growth process that gives priority to the reduction of poverty, unemployment and inequity, aims at the satisfaction of basic minimum human needs, and stresses self-reliance and the participation of all the people—especially those with the lowest standards of living.

In these alternative approaches and strategies, there is a need to emphasize and not merely recognize the agrarian core of most Third World countries in general, and in the Asian and Pacific region in particular. Rural development must be considered both an integral part and the driving force of the entire development process. Thus rural development should not be considered as a "sectoral" activity; rather it should be seen as the multi-sectoral core of a broader development process. Simple logic forces the countries in the region to give more weight to agrarian issues; the sheer mass of agriculture in most economies of the region and the need to match growing population with more jobs. There is no denying that poverty and its associated problems are concentrated within the rural society in the region.

Rural development is not only a question of reversing an urban bias in development and planning and implementation, and allocating more financial and technical resources to rural areas. The desired transformation of agrarian society, in most cases, requires a change of the structure of political and economic power, both local and national. The nature and direction of these changes will clearly vary according to national history, values and traditions. The pace of the transformation process will also vary according to the degree of resistance encountered as well as to the degree of perception by national leaders that the national destiny cannot be achieved realistically without the structure transformations.

In this context, it was agreed that the pace or specific changes to be brought about could not be prescribed. Indeed, realism dictated that deliberations on development should be framed within the current national political and economic structures but leavened by the belief that there is a growing consciousness at national levels that structural changes are both necessary and desirable.

Objectives for Alternative Strategies

Alternative strategies should be focused on an integrated set of objectives whose overall purpose is to increase the quality of life of all persons but particularly of those who have not benefited equally from development. It is recognized that most of these objectives have characterized past national and international development strategies, which failed to achieve them. What is different is the relative priorities given and the relationship among the objectives, leading to a new way of looking at development and toward the design of alternative strategies with a greater chance of success. The fundamental objectives include:

1. *Increasing productivity and growth in output* : It is recognized that without orderly growth other objectives cannot be achieved, and that growth itself is not an issue; rather it is the nature of the growth which is critical.
2. *Equitable distribution of the benefits of development* : The emphasis here is not on the ultimate distribution of benefits, but on distribution as a key ingredient in generating a pattern of growth that is broad-based.
3. *Provision of basic human needs to the whole population*: As a fundamental part of the process of equitable distribution a minimum level of basic human needs, i. e. food, clothing, shelter, health and education, must be provided for all.
4. *Provision of gainful employment* : This recognizes that productive and gainful employment is both an end and a

means and is, in fact, one of the most basic human needs, as well as a critical factor for development of human resources.

5. *Effective participation in decision-making, especially by the poor* : This recognizes that participation is both an end, desirable in itself as a human need, and an essential means to assuring the pace and direction of development. It is recognized that participation in its essence must be in terms of decision making, i. e. determining the course of one's own actions, implementation and monitoring.
6. *Self-reliance* : While recognizing that neither autarky nor other means of setting off national or local development apart from other forces are desirable, development should be self-sustaining, make use of available resources especially the resources provided by the people themselves, and have a minimum dependence on external factors. Without this, development cannot be self-directed. What is important is interdependence at various levels within the nation and outside on a consultative basis rather than on a compulsive basis.
7. *Maintenance of an ecological balance* : Without prejudice to the need for use of resources to generate growth, there should be, both as an end and a means, an effort to maintain a fair ecological balance in the use and control of available resources. Without this, development of the present will be at the cost of development in the future.

These objectives are seen as an *integrated package.* They represent a specific development style, which is not imitative and is indigenous to every society. The key to this style of alternative strategies is to minimize potential contradictions within and among the objectives, and use the achievement of one to further the achievement of the other. It should be noted that the central objective, binding the others, is the effective participation of the people in development in terms of decision-making, providing the resources for development, and in determining how the benefits of development will be shared.

Without this participation, the other objectives cannot be achieved.

Necessary-conditions for Formulating Alternative Strategies

Achievement of these objectives and the formulation of strategies for this requires that several conditions be met. In a broad sense, a major necessary-condition is the existence of an international political and economic order which does not work against the objectives. At the national level there must be (1) a political will to make the decisions necessary to achieve the objectives; (2) there must be effective national planning and the elaboration of consistent policies designed to achieve the objectives; (3) there must be equitable access to productive assets; and (4) there must be organizational structures at the local level. It should be noted that the objectives proposed do not differ greatly from those which have been espoused in the past. However, there must be a firmer national consciousness of the nature and implications of them.

The Role of Local-level Planning and Development in the Alternative Strategies

The importance of local-level planning and development derives from the central objective of people's participation. The people are not an abstract; they live and work in villages and towns. Engaging them in the development process means involving them at the level where they are. Interventions by the public sector are most meaningful at that level. The inequities of existing economic and political relations find concrete expression there. While this might appear to be a truism, its validity is often ignored when development is seen as policies determined at national levels and when actions are seen in terms of the government programmes and their agents rather than in terms of the recipients and intended beneficiaries.

In this context, the key to an alternative strategy is its focus on the local level. The concern is with how to direct and control development through pluralistic organizations of the people themselves, while providing, at the same time a form of national and regional consistency which permits effective interaction with the international level.

Elaboration of strategies, therefore, requires attention to

two distinct problems : the organization of local level planning and development itself and the linkages between this level and higher levels.

LOCAL LEVEL PLANNING

Rationale of Local Level Planning

The rationale for local level planning and development may be stated to be fourfold : (a) better perception of basic issues at the 'grassroots' level with specific attention to the rural poor, and formulation and execution of programmes and projects for their solution ; (b) better exploitation of the growth potentials of an area for improving productivity and increasing production ; (c) opportunities for direct participation by the local population, particularly the poor majority in development decision-making directly at the lowest (homogenous groups, hamlet, village) level and between representatives at higher, sub-district, district level, also opportunities for participation of the target groups concerned as an end in itself—for inculcation of democratic values and education in rights and obligations of citizenship as well as developmental education, and (d) mobilization of local resources, mainly surplus labour.

Definition of Local Level and Local Plan Area

Local level, on the basis of the rationale for direct participation by the majority in the development process, has to be the smallest unit of homogenous small groups, a hamlet or a compact (not necessarily homogenous) village or community which will be conducive to direct participation by, as distinct from indirect representation of, the majority, particularly the underprivileged, in programmes and tasks of their own development. 'Local' level development as understood respectively in terms of the politico-administrative system in a given country would mean the lowest echelon of the local government as well as of the administrative hierarchy. If the latter two do not converge, as they need not necessarily do so, then three discrete levels within the local level may be identified, namely, the lowest level of homogenous small groups of members formed around productive activities or common interests, the local government of the village or multi-village level, and the politico-administrative unit at the sub-district and/or district levels.

The area of the largest of the three units within the 'local level' may be considered as the viable unit for purposes of local 'area' planning.

Since the jurisdiction of the lowest administrative unit often happens to be larger than the unit of the local government the former becomes usually the local level 'area' for purposes of planning :

A local level 'plan' would thus consist of distinct programmes and activities to be formulated and implemented at the three levels within the local 'area'. In the plan 'area' the following conditions should be met :

- Effective direct and/or indirect participation by the majority ;
- Allocation of adequate resources from the centre/regional level of government ;
- Generation and mobilization of local resources,
- Provision of technical services by the central/regional government ;
- Area and population should be adequate and large enough for viable 'area planning';
- Existence of plural homogenous groups and people's institutions, including local government units, *to act as countervailing forces.*

Task of Local Level Planning and Development

The tasks of planning at the local 'area' level as defined above would comprise the following :

(a) to interlink and coordinate the planning and development activities at the levels of the homogenous small groups, the local government and the administrative unit, and formulate a well integrated local 'area' plan;

(b) to link the plans and policies for the local 'area' with regional and national policies and plans;

(c) to mobilize resources at local and from supra-local levels in order to match the planned programmes and activities with the available resources;

(d) to implement, monitor and evaluate the development plan and programmes in terms of outputs, effects and impact, with particular reference to the target groups.

Necessary Conditions for Local Level Planning and Development

There are five pre-requisites for meaningful local planning and development, namely, resources, institutions, participation by the under-privileged majority in decision-making, consultative planning process and linkages between local and national/regional planning processes

Resources

Resources must be made available at the local level beginning from the local politico-administrative unit down to the village and group level. A decentralized budgeting system, a system for putting in necessary provision of technical inputs and expertise must be integrated with the local planning process. It would be preferable if the local organizations i.e., local government bodies or regional politico-administrative units are given the power and the mechanism for generating their own resources through taxation, for instance. But if this is not possible because of political constraints, then, at least the funds should be placed at the disposal of localities on the explicit understanding that these are collected in one form or another, from the people of the area and are being returned by the central or regional government to the localities *as a matter of right*. This is an important point. If it is doled out as a 'grant' the possibilities of wastage are more. Further, popular involvement depends to a large extent on the pride of ownership.

Community participation in, and complete devolution of authority and responsibility to local communities for, acquisition of local resources, mainly land, for purposes of community projects is an important area where the community can meet costs of compensation to those who would lose land or resources by a system of progressive taxation of those who would or are likely to gain from the completion of the projects, such as roads and irrigation channels, etc.

Institutions

Second, there must be viable institutions, including adequate politico-administrative structures at the local level.

Besides "political will" at the highest level and preferably at all levels of the political leadership in a country, which is willing to accept some calculated risks involved in any development strategy, there is need for devolution of political power to local government units as well as decentralization of adequate authority, including financial and technical resources, to the lowest units of administration. The aim should be to provide autonomy to local areas in matters of decision-making within the framework of national policy and plan, and not mere delegation of authority or deconcentration of administrative units.*

In many countries the traditional administrative structure that still exists, particularly at the local level, is not only undermanned but is attuned to the maintenance of law and order and revenue collection. It would be necessary to reorientate the structure as well as the staff for developmental activities. Above all, there will have to be an attitudinal change in the bureaucratic staff leading to greater empathy to the needs of the poor majority and also willingness to accept consultations with and guidance from the people. Such attitudinal changes as well as better performance by the bureaucracy can be most effectively brought about by a built-in-system of "performance appraisal by objectives", such as described in one of the discussion papers,** in the public service. Also, as already indicated, a healthy tradition of public accountability should be instituted and carefully nurtured by building up informal pressure groups at the local level. Effective participation by the citizens at the grassroots levels in monitoring and evaluation of development programmes and projects in open assemblies,

* For instance, for such autonomy to be meaningful both administrative and technical staff seconded from central/regional civil services will have to be put in most cases, under the single control of the local authority, as opposed to the conventional system of dual control of the line department, as well as the local authority. Such a system of secondment will have implications for the central/regional/civil service systems. Creation of a separate civil service system for local authorities in most cases may be neither a feasible nor a desirable proposition.

** Chapter 2

coupled with "performance appraisal" of those responsible for the execution of those projects will have a salutary effect on both the bureaucracy as well as local politicians.

There is need for building up two way communication and information channels between the local authorities and the populace. Such communication and sharing of information is necessary for greater perception of people's needs on the one hand, and for better participation in decision-making by the majority on the other. It may also be worth pointing out here that the approach to development envisaged in the alternative strategy for local level planning has its own implications for the information system. Extensive data would be required for various levels and institutions with appropriate linkages. Countries would, therefore, need to consider carefully the type of information required to meet local needs.

People's Participation

Third, people participate in what is theirs. There must be something on the ground to be involved in. There are, of course, local organizations in existence, but these tend to be dominated by the local elite. This is one major reason why the poor and the underprivileged have not been able to participate in the developmental process. It is, therefore, necessary that a system of plural local organizations be built up outside the formalised and traditional framework, including local government. These countervailing organizations may be production groups, voluntary associations, farmers' groups, youth clubs, women's associations etc., depending on the local conditions. Interesting experiences in this regard by voluntary agencies, such as the Rural Library and Rural Science movements in Kerala, India, were taken note of. These organizations are to be encouraged by the national government. These must have resources, both financial and technical. There may be conflicts between the new organizations and the existing ones. But realistically there cannot be change without bringing into surface the already existing conflicts. The solution is that especially national governments, and also non-governmental bodies, research institutes, etc. give sufficient support to the new organizations. Hopefully, in the long run it would be realized at the local level that it is to the advantage of all concerned to co-operate.

But these organizations at the lowest level have still to be linked to the sub-regional, regional and national levels of operations. It depends very much on the political constraints of each society. In a country with a regular elective process, genuine representation and public accountability of politicians, decentralization of budget and planning to local politico-administrative units, as in Sri Lanka, may work. But, elsewhere some other instruments and administrative structures may be found. For instance, regional/local development corporations, as suggested in one of the background papers, may have to be set up with the co-operation of public leaders of that region or locality, who command the respect of the majority of the people. Yet in other cases, a viable alternative may be to use non-governmental voluntary organizations, academic or research institutes for local and regional planning. These linkages and organizations will have to be designed according to local conditions in each society.

Consultative Planning Process

Fourth, is the process of development itself. This can be divided into two parts, one dealing with the activities of generating income and the other, providing the necessary basic social services.

The first part is related to the productive process itself. The basic components of an activity that generates income are (a) pre-investment investigations, (b) investment, (c) production operation, (d) marketing of output, and (e) sharing of benefits generated by this activity. It becomes obvious that these processes cannot be neatly divided. Inputs from outside the local production unit are needed, like technical support for investigation, and actual inputs, i.e. fertilizer, water for irrigation, etc. Similarly, marketing cannot be left to the group itself which cannot compete with outside organized forces. Each activity has to be analysed for locating the decision-making levels and points. Then it becomes a consultative process between various levels with forward and backward linkages so that eventually, some rational decision may emerge.

The second part of the development process has to do with the basic social services. If people are so poor that their needs cannot be satisfied, generating income activities cannot be undertaken. For meeting the basic needs, which are the inalienable right of every human being, resources have to be reallocated and/or transferred from the richer regions to the poorer regions, from the richer sectors of population to the poorer sectors of population. The system of resource transfer, i.e. the mechanism thereof, has to be politically acceptable and fitted to local conditions and resource endowment including its spread.

Linkages between Local and Other Levels of Decision-making

Fifth, a major implication of the consultative planning approach for national policy is that the national goal and local plans have to be consistent with each other. If self-sufficiency in food grains is a national goal, production of grains would have to be given priority in resource allocation at the local level. On the other hand there would also be the need to look at the development process from the other end of the telescope, that is to say, from the perspective of local development. Such a shift in perspective or approach will have implications for development policies and programmes at the central/regional levels. Some of the implications are quite obvious. They are not new. It is only a question of shift in emphasis. For instance, there would be a shift toward labour intensive technology with the attendant problems of productivity and organization, a redirection from heavy industry to agriculture and light industry with all the infrastructure development involved i.e. irrigation and extension services, comparatively greater reliance on smaller and medium projects rather than large ones. There will be a shift from the so-called directly productive kind of investment to the social service sector, i.e. health and education. Some of this shift may be helpful for growth in the conventional sense, some may retard growth. The final decision for a balance is a political one, and as such, it will have to be decided through the existing political process. Some improvement therefore may be marginal, some may be significant, depending on the existing socio-political structure.

Local development approach, however, if pursued logically

will result in clear and sometimes far reaching shifts in the national development strategy, sectoral as well as regional. For example, the educational policy and training programmes, formal, informal and non-formal, may need a thorough overhaul in order to generate initiative instead of passivity among the youth, functional knowledge and skills instead of theoretical bookish knowledge, a sense of social responsibility instead of purely individual profit motivation, a sense of balanced development instead of urban bias, inculcation of an indigenous value system instead of alien values, etc.

All training programmes will also have to be reoriented afresh and their curricula revised in order to train a new breed of "barefoot" technicians who would have their origins in the rural areas and will be willing to live and work in those areas. Also training programmes for government functionaries and people's nominees at the local level in simple techniques of project identification, formulation, appraisal and implementation are necessary for building up local planning capabilities.

The whole system of subsidies and incentives for agricultural and industrial development, to take another example, will have to be looked at from the point of view of their impact on the majority and the need to provide them with a basic minimum of wage goods. Subsidies ostensibly meant for the poor seldom benefit them, for more often than not the target groups are not in a position to avail themselves of subsidized services. It may be necessary therefore to reexamine the indirect "hit-or-miss" strategy of subsidies and devise new means of transferring the resources thus saved directly to benefit the target groups in the form of additional incomes or access to public and essential services through employment. An innovative scheme for such transfer through "labour certificates" has been described at some length in one of the background papers. The proposed scheme of "labour certificates" should also open up possibilities of generating employment for the rural underemployed and unemployed and the creation of a social infrastructure as well as increase in production through numerous productive activities, on a self-sustaining basis. The scheme, or at least the essential concept involved in the scheme, may be worth examination by national planners and administrators for adaption as a strategy or instrument for local development.

In this context, however, it is to be clearly understood that what is being advocated is not reappropriation or redistribution but a rational reallocation of resources. However, poor a society is, there are vast areas of potential improvement in terms of resource allocation and utilization, as in many countries a great deal of resources are being used for the wrong purposes. It is, therefore, a matter of redirected resource use consonant to the objectives set forth for a development that is focused on the people. This will require, besides the political will of sovereign national governments, some administrative re-organization or reallocation which would make it possible to build up the systems of linkages identified above.

Issues for Further Research and Study

Local level planning is a discipline to which very little attention has been given so far by development planners, administrators, or policy makers. It still remains to a large extent an un-explored terrain.

Hence several important issues were raised at the meeting. Most of these, such, as relating to methodology of planning at the local level were deferred, as already indicated, for discussion at the next forthcoming ESCAP meeting on "Local level Planning for Integrated Rural Development".

Some of the important issues that would need more analysis and research and experimentation may be briefly indicated as follows :

1. Action research for systematic collection of empirical evidences as local organizations outside the formal government structure;
2. Experimental projects on local level planning, imple-mentation, monitoring and evaluation;
3. Linkages between the three levels within the local 'area' level, i.e. homogenous small groups, local government and lowest politico-administrative unit, from the point of both the planning process and people's participation in decision-making at those levels;
4. Politico-administrative structures and civil service systems at the local level and their linkages with the

bureaucratic hierarchy at higher levels, particularly with a view to ensuring autonomy and effective decentralization of authority at the local level.

5. Two way linkages with, and implications of local level planning and development for, central and intermediate level government policies, plans and programmes; or the framework of planning looked at from the bottom as distinct from the top;
6. Special attention to new information needs, particularly acquisition and systematic flow of information from local level to higher levels and vice-versa.

These are some of the broad issues that need further analysis and investigation. This report has indications and suggestions of several other issues that may be taken up by the ESCAP meeting referred to, as well as for incorporation in the work programme of UNAPDI and other institutions associated with the "Interagency Committee on Integrated Rural Development in the ESCAP Region".

ANNEX : LIST OF PARTICIPANTS

Javed Azfar
Planning Commission
Government of Pakistan
Secretariat Block A
Islamabad, Pakistan

A.Z.M. Obaidullah Khan
Secretary
Ministry of Agriculture and Forests
Government of the People's Republic of Bangladesh
Dacca, Bangladesh

M. Sadiq Malik
Executive President
Rural Development Foundation of Pakistan
Islamabad, Pakistan

K.J. Ratnam
Director
Centre for Policy Research
University Sains Malaysia
Penang, Malaysia

John Samy
Senior Planning Officer
Central Planning Office
Government of Fiji
Suva, Fiji

Phaichitr Uathaweekul
Rector
National Institute of Development Administration (NIDA)
Klong Chan, Bangkapi
Bangkok, Thailand

Wickrema Weerasooriya
Secretary
Ministry of Plan Implementation
Government of the Democratic Socialist Republic
of Sri Lanka
Colombo, SRI Lanka

Somsakdi Xuto
National Institute of Development Administration (NIDA)
Klong Chan, Bangkapi
Bangkok, Thailand

REPRESENTATIVES OF THE UNITED NATIONS BODIES AND SPECIALIZED AGENCIES

ESCAP

Sultan Z. Khan
Chief
Agriculture Division

P. Laplante
Senior Officer
Agriculture Division

B.P.W.M. van Dusseldorp
Consultant
Agriculture Division

L.R. Khan
IRD Unit
Office of the Executive Secretary

A. Hashem
Officer
Development Planning Division

Salima Omer
Population and Social Affairs Division

R.H. Sherif
Chief
Compilation and Analysis Section
Statistics Division

S.G. Tiwari
Regional Adviser on National Accounts
Statistics Division

United Nations Headquarters

John Mathiason
Secretary
Rural Development Task Force

UNICEF

A. Welle
Regional Food and Nutrition Officer

UNDP

R. England
Assistant Regional Representative

ILO

M.A. Khan
Regional Adviser on Rural Institutions

FAO

G.C. Clarke
Small Farmer Development Officer

B.P. Dhital
Programme Officer

UNESCO

Yogesh Aatal
Regional Social Science Adviser

Thamrong Buasri
Specialist in Curriculum Development

WHO

M. Sathianathan

ILO/ARTEP

K.N. Raj
Chief Technical Adviser

S.M. Naseem
Development Economist

Wilbert Goonaratne
Rural Development Expert

H. Poot
Associate Expert in Manpower Planning and Statistics

UNAPDI

Vinyu Vichit-Vadakan
Director

Ram C. Malhotra
Deputy Director

D.V. Ramana
Development Economist

Ram P. Yadav
Consultant

Narongchai Akrasanee
Consultant

Vijita B. De Silva
Development Planner (Development Information)

OBSERVERS

Hiran Diaz
Associate Professor in Rural Development Planning
Asian Institute of Technology
Bangkok, Thailand

Kurt Dopfer
Professor of Economics
Swiss National Fund for Scientific Research
Berne, Switzerland

Bruno Knall
Director
South Asia Institute
Heidelberg University
Federal Republic of Germany

R.H. Retzlaff
Associate
Agriculture Development Council
Bangkok

Richard Vokes
Assistant Professor
Rural Development Planning
Asian Institute of Technology
Bangkok, Thailand

APPENDIX

NOTE ON UNAPDI RESEARCH PROJECT ON LOCAL LEVEL PLANNING AND DEVELOPMENT

UNAPDI has a research project on Local Level Development with Focus on People's Participation, approved by the UNDP for its programme support from 1 July 1979 to 30 June 1980. The project is intended for predominantly agrarian economies and thus mainly, although not exclusively, for the least developed and land-locked countries in the ESCAP region UNAPDI has initiated the project in Nepal, Bangladesh, Maldives, Sri Lanka and Thailand. UNAPDI has been collaborating with the Inter-Agency Committee and Task Force for integrated Rural Development in the ESCAP region, the ESCAP Secretariat and ARTEP/ILO, both in the conception and implementation of the country studies, to the extent possible.

Background

UNAPDI and ESCAP jointly convened the following two meetings in October-November 1978 :

1. UNAPDI Consultative Meeting on "Alternative Strategies for Development focusing on Local Level Planning and Development", Bangkok, 31 October to 4 November 1978; and
2. ESCAP Expert Group Meeting on "Local Level Planning for Integrated Rural Development", Bangkok, 6-10 November 1978.

The objectives of the first meeting were :

(a) to systematically identify elements and clarify concepts in outlining alternative (or new) strategies or approaches to development; and

(b) to delineate the role and scope of local level planning in the context of the new approaches to development.

A detailed examination of issues and techniques involved in local level planning based on a review of experiences in selected countries in the region was made at the second meeting.

Based on the recommendations of the two meetings, UNAPDI, in consultation with the Inter-Agency Task Force on Integrated Rural Development, ESCAP and ARTEP/ILO and informal consultations with countries and national institutions in the region, drew up a project proposal on "Local Level Development" for UNDP programme support. The UNDP support became available in mid-1979. The project proposal as approved by the UNDP envisages action-oriented empirical research studies in at least four countries, with priority being given to the least developed and land-locked countries in the ESCAP region.

Purpose

In the face of widespread disillusionment with growth-oriented top-down approaches to development, the broad purpose of the project may be stated as follows :

— to identify crucial elements and issues for designing an appropriate (alternative) strategy for national/rural development with focus on people's meaningful participation. Since the majority of the people in predominantly agrarian economies live in rural areas and in local communities, the focus has been on local level/rural development. Again, such a strategy has to be developed in the specific socio-economic-political-cultural context of a given country.

— to review linkages with, and implications of, *local* level planning and development for *central* and *intermediate* level government policies, plans and programmes; in other words, to review the framework of the process of planning looked at from the bottom, as distinct from the top.

Objectivies

The objectives of the research project are :

1. To evaluate the recent past and ongoing efforts on local level planning and development on the basis of the following criteria :

(i) productivity and equity—basic needs oriented;
(ii) employment generation;
(iii) genuine people's participation in decision-making;
(iv) self reliance, or optimum mobilisation of local resources, including surplus labour;
(v) eco-development; and
(vi) other national priorities, if any;

(These criteria are derived from the objectives of an alternative strategy for national/integrated rural development, arrived at the UNAPDI Consultative Meeting held on 31 October —4 November 1978).

2. To come up with a design and recommendations regarding appropriate institutions and built-in mechanisms for participatory development at the local (village or community, thana, district) levels in order to achieve the objectives (criteria) given under 1 above in the socio-economic-political-cultural context of a given country.

3. To prepare draft village/community/district development plans for possible implementation by the authorities concerned.

The project is basically action-oriented, in the sense that the objective is to come up with viable and practical methodologies including recommendations regarding necessary institutional and procedural changes etc. given the existing socio-economic structures of a society, with a view to involving the majority of the people, particularly the poor at the 'grass roots' level in a gainful, equitable and self-sustaining manner.

Nepal

The project in Nepal has been started since mid-1979 under the auspices of the National Planning Commission and the Ministry of Home and Panchayats and in collaboration with national training and research institutions like the Centre for Economic Development and Administration (CEDA), the Agriculture Projects Services Centre (APROSC), the Panchayat Training Institute, and development agencies such as the

Agriculture Development Bank (ADB/Nepal). UNICEF and Ford Foundation are also participating in the Nepal project with substantial contributions of funds. The Ford Foundation has also made available the services of its consultant to CEDA, Dr. Pradipto Roy, on a full-time basis.

The project has a Steering Committee composed of the participating agencies and institutions to design and implement the research project. The Steering Committee has decided to select a minimum of six districts from different regions of Nepal, and to send multi-disciplinary survey teams composed of at least two professional staff members drawn from the participating institutions. The teams will be spending two months in their respective districts collecting data and interviewing various sections of the local population, such as the landless and marginal farmers, small farmers, middle and larger farmers, artisans and other non-farmers, and finally, professionals, such as government workers, local leaders including women leaders, with a view to assessing their various priorities and needs for development at the village or community and the district levels, and also their perceptions regarding institutional structures and mechanisms necessary for people's effective participation in development decision-making and equitable benefit sharing.

The field surveys are expected to begin in late February or early March 1980, prior to which the Steering Committee and Task Force are expected to complete discussions on, and the formulation of, substantive issues, such as objectives of development at the local level, criteria for evaluation of recent and ongoing programmes in local level/rural development, institutions and mechanisms for effective participation by the majority of the people at the local level, methodology of field research, training of field investigators and collection of available data from secondary sources regarding the six districts.

Each of the district survey teams will be expected to prepare its report focusing on the findings regarding the issues mentioned above, and the methodology adopted in arriving at its recommendations regarding reforms in institutional structures and mechanisms necessary for effective people's participation in local level development, and ensuring its linkages with (subnational) area and macro-level planning and development programmes. Another output will be a draft district development plan, more with a view to supplementing and improving the plans already submitted by each district's administration to the Planning Commission for the Sixth Five Year Plan (1980 to 1985).

The survey reports are expected to be discussed at district level seminars attended by district-wide development workers at all levels—knowledgeable, political, bureaucractic, private and voluntary, persons drawn from all sections of the district.

Based on the comments made in the seminars, the district development reports and plans will be revised. Based again on these revised reports, the Steering Committee and its Task Force will be expected to prepare guidelines or a draft n anual for local level development. These guidelines, as well as the six draft district development plans will be finally submitted to His Majesty's Government, through the Ministry of Home and Panchayats and the National Planning Commission.

It is hoped that all the tasks will be completed by 30 June 1980 in time for the beginning of Nepal's Sixth Plan (1980 to 1985) on 1 July 1980.

(It is worthwhile mentioning that the National Planning Commission has requested the Steering Committee to feed it with all information and research findings, however tentative, regarding district development planning as soon as it becomes available, without waiting for the final submission).

Bangladesh

The research project has been initiated in Bangladesh with the formal appointment of a project Coordinating Committee composed of senior officers from Integrated Rural Development Department, Ministry of Local Government, Rural Development and Cooperatives; Ministry of Agriculture and Forests; Bangladesh Bank; Bangladesh Institute of Development Studies (BIDS) Bangladesh; Academy for Rural Development (BARD/Bogra); Civil Officers' Training Academy (COTA); and a few voluntary agencies such as Gonoswasthya Kendra and Bangladesh Rural Active Committee (BRAC).

In view of the numerous innovative experiments in local level/rural development being carried out in recent years in the country, the research project in Bangladesh will be limited to an in-depth evaluation of selected ongoing projects and programmes rather than on designing any new strategy for local level/rural development.

It may be pointed out, however, that although the scope of the project will be limited to an evaluation of selected ongoing projects, it will have to come up with recommendations and possibly guidelines for local level development with people's participation on the basis of the various evaluation studies.

Maldives

The research project in the Maldives is being carried out concurrently with, and as an integral part of, the Atoll Development Project, being undertaken at the request of the Maldives Government by the Inter-Agency Committee on Integrated Rural Development. The project started in Alifu atoll in November 1979 under the guidance of an inter-departmental committee on which are represented the National Planning Agency, Ministries of Provincial Affairs, Agriculture, Fisheries, Education, Health, Transport and Communications and Atoll Development Advisory Board. The project is headed by a short-term Consultant assisted by an FAO expert and a UNAPDI Consultant. The World Bank Economic Advisor to the National Planning Agency has also been associated with the project. The project is expected to be completed by April 1980.

Sri Lanka

Originally the project in Sri Lanka was also expected to be carried out as an integral part of an Inter-agency IRD Committee project on district planning, as in the case of the Maldives. But delays regarding finalisation of various issues regarding the project led UNAPDI to redesign the project along lines similar to the Bangladesh project. Accordingly discussions were held with government authorities in November 1979, and a project has been initiated under the aegis of a Steering Committee under the chairmanship of the Secretary of the Ministry of Plan Implementation and composed of the Secretary, Ministry of Rural Development, Director, National Planning; Director, Marga Institute; Director, Agrarian Research and Training Institute (ARTI); and Representatives of non-governmental Voluntary Organisations such as Sarvodaya, Redd Barna and

Community Development Services. A Director of the Ministry of Plan Implementation has been also appointed to act as coordinator of the research project.

The Steering Committee in its first two meetings held in January 1980 decided to concentrate on the grass-root level of village development, rather than the district level and also decided to make analytical/evaluation studies of all ongoing projects at village level undertaken by Marga, ARTI, Ministry of Rural Development. Sarvodaya (NGO) and the Norwegian programme, for consideration by the Steering Committee.

The deadline for the project, originally set at 31 March 1980, has been extended to 30 April 1980.

Thailand

In Thailand, unlike in the other four cases, the project will consist of evaluation studies of three projects, located in Sarapee, Maeklong (Samut sakorn) and Yesothorn districts undertaken by three distinguished social scientists from a national research institution, namely, Thai Khadi Research Institute, Thammasat University, Bangkok.

The Sarapee project of the Community Development Department has been selected by the government as the model of an integrated rural development project for replication throughout the country. The Maeklong project was originally started by three universities, Thammasat, Chulalongkorn and Kasetsart.

The research studies are expected to be completed in mid-May 1980.

Collaboration with ESCAP and ARTEP/ILO

Both ESCAP and ARTEP/ILO have also undertaken projects in local level planning. Hence UNAPDI has been keen in maintaining close collaboration with both in order to mutually support each other in their respective country-based studies. ARTEP has generously offered to provide the services of its consultant, who designed and carried out household surveys and field research in its Bangladesh project, for the UNAPDI project in Nepal. The offer has been conveyed to the Steering Committee in Nepal.

ESCAP's project envisages reviewing current approaches to

local level planning in selected rural areas in five countries and "coming out with detailed suggestions for strengthening the planning process in order to achieve better results for the benefit of the rural population concerned". The findings of these studies are expected to be submitted to national workshops for their consideration by the end of 1980. UNAPDI has suggested to ESCAP (Agriculture Division) to make use of its machinery of Steering/Coordinating Committees in Bangladesh, Nepal and Sri Lanka, since these happen to be countries common to both the ESCAP and UNAPDI projects. The objectives and purposes of both the projects are similar, namely to draw up guidelines for local level planning and development in the context of multi-level planning. The methodologies however are somewhat different. The UNAPDI project is based on field research, focused on assessing people's perception of local development needs and priorities and institutions and mechanisms for ensuring people's effective participation in decision-making as well as evaluation of recent and ongoing efforts in local level development, including current planning systems at the local level. ESCAP project will be mainly based on review of existing systems of planning and would not involve field research.

The two studies, therefore, can be mutually reinforcing and complementary. Another common feature of both the projects is that the draft guidelines are envisaged to be discussed in national seminars (also in district seminars at district level in the case of the UNAPDI project in Nepal) in both the projects. But in view of the fact that UNAPDI will cease to exist after 30 June 1980, it would not be possible to organise national seminars in all the five countries. (At best district level seminars in some out of the 6 districts in Nepal may be held, but we also doubt the feasibility of holding such district level seminars because of the national referendum in May-June 1980). Hence the national seminars envisaged in the ESCAP project on local level planning may serve the purpose of the UNAPDI project also. (This will be necessary only in case the UNAPDI project is not continued by APDC).

Future of the Project

This will however be only the end of Phase 1 of the project, that is to say, as far as UNAPDI's responsibility is concerned under the current UNDP approved project. But it is expected that as far as UNDP and the countries are concerned, the project should continue for new phases to undertake tasks such as the following that would have remained unfinished at the end of the current phase of the project, i.e. by 30 June 1980 :

1. Continuation of the survey of the remaining districts (such as in Maldives, Nepal and Sri Lanka)—task to be undertaken by national agencies and institutions with the assistance of their own staff members;
2. Refining of the guidelines for local level development on the basis of further insights and experience gained with (1);
3. Training of staff of the participating institutions to undertake (1);
4. Creation of planning cells at the district level and training of their staff in district level planning and development;
5. Training of village level planners (officials/elected representatives) in village/community level development;
6. Setting up institutional structures and mechanisms for people's participation and testing their efficacy;
7. Monitoring and evaluation of all the above measures and preparation of manuals for village, district, subnational (regional, area) and national (macro) level planning, implementation, monitoring and evaluation and equitable sharing of the benefits of development, and soon.

Bangkok
18 February 1980

U.N. Asian and Pacific Development Institute

PART II

LOCAL LEVEL PLANNING

INTRODUCTION

THIS PART is meant to reflect the current state of the art of local-level planning for rural development in the countries of the ESCAP region and to suggest ways of improving the national capacity for achieving desired results along these lines.

The area covered by "local-level" planning usually coincides with the lowest administrative echelon (district or sub-district) and planning carried out in close communication with the villages, where the rural population can be effectively involved in the planning and implementation process of rural development. As indicated in chapter eight, local-level planning is a phenomenon of relatively recent origin coinciding with the realization by Governments that the involvement of the rural people, and particularly the rural poor majority, in their development is of the utmost importance. Its rationale is based upon its role in facilitating decision-making by the rural poor, in formulating and implementing development projects and activities more acutely attuned to their requirements and in mobilizing and allocating resources to exploit fully the growth potential of local areas.

The Expert Group meeting formulated a series of suggestions for follow-up action ; and ESCAP, in co-operation with the other agencies concerned, intends to provide full support to member countries wishing to improve their capacity for local-level planning in rural areas.

Sincere appreciation is due to the Government of Australia for the generous grant which made the Expert Group Meeting possible.

EIGHT

REPORT OF THE EXPERT GROUP MEETING

ORGANIZATION OF THE MEETING

THE Expert Group Meeting on Local-level Planning for Integrated Rural Development, convened by the Economic and Social Commission for Asia and the Pacific (ESCAP) in co-operation with the United Nations Asian and Pacific Development Institute (UNAPDI) and the Interagency Committee on Integrated Rural Development for the ESCAP Region, was held at Bangkok from 6 to 10 November 1978.

Attendance

The meeting was attended by experts from Bangladesh, the Federal Republic of Germany, India, Indonesia, Malaysia, Nepal, Pakistan, the Philippines, Sri Lanka, and the Union of Soviet Socialist Republics.

The meeting was also attended by officials from United Nations Headquarters and ESCAP. The following United Nations bodies were represented: United Nations Children's Fund, United Nations Development Programme and United Nations Environment Programme.

Representatives of the following specialized agencies attended: International Labour Organisation, Food and Agriculture Organization of the United Nations, United Nations Educational, Scientific and Cultural Organization and World Health Organization.

Representatives of the following organizations were also

present: UNAPDI, Development Training and Communication Planning and Agricultural Development Council.

Opening Address

The Meeting was inaugurated by the Executive Secretary of ESCAP. In his opening statement, he welcomed the participants and informed them that the Expert Group Meeting was the second of two meetings on local-level planning which had been planned jointly by ESCAP and UNAPDI, of which the first[1], had been expected to put the emphasis on the broad policy aspects whereas the second was expected to focus on the operational aspects of local-level planning for integrated rural development. The Meeting was also linked with the efforts of the Interagency Committee on Integrated Rural Development for the ESCAP Region to support national programmes of rural development.

The Executive Secretary indicated that member countries of the ESCAP region were very much concerned with the problem of involving the rural masses more effectively in the process of development. That required a broad-based, decentralized approach taking full advantage of the human and natural resources available in the rural areas. In that context, local-level planning, with the framework of national strategies, became imperative in order to achieve a coherent pattern of rural development on a sustained basis.

He said that the purpose of the Meeting was to review the current practice of local-level planning in selected countries of the region as a basis for suggesting ways of improving national capacity for achieving desired results along those lines. He hoped that the Expert Group could also suggest a joint programme of follow-up action, to be conducted in the spirit of technical co-operation among developing countries (TCDC) with the support of the international community. He expressed special thanks to the Government of Australia for its financial support of the Meeting and to the Government of the Netherlands for making available a special consultant.

Election of Officers

Mr. Shoaib Sultan Khan (Pakistan) was elected Chairman of

the Meeting. Mr. L. J. Fredericks (Malaysia) was designated as rapporteur.

Adoption of the Agenda

The following agenda was adopted:

1. Election of officers;
2. Adoption of the agenda;
3. Review of current practice of local-level planning in selected countries, in the light of the guidelines formulated at the previous meeting (31 October—4 November 1978);
4. Examination of the major constraints experienced by member countries in this field and formulation of guidelines for improving the national capacity to achieve desired results along these lines;
5. Suggestions for a regional programme of action to support national efforts in that direction (including TCDC as well as bilateral and multilateral support)
6. Adoption of the report.

REVIEW OF CURRENT PRACTICE

The experts, referring to the country papers (see chapter eleven), reviewed the current practice of local-level planning for rural development in their respective countries. The highlights of the discussion may be summarized as follows.

Scope of Local-Level Planning

Local-level planning for integrated rural development was a phenomenon of relatively recent origin coinciding with the realization by Governments that the involvement of the rural people, and particularly the rural poor majority, in their development was of utmost significance. In general terms, the rationale of local-level planning was based upon its role in facilitating decision-making by the rural poor, in formulating and implementing development projects and activities more acutely attuned to their requirements and in mobilizing and allocating resources to exploit fully the growth potential of local areas.

While basic agreement was evident regarding the rationale and need for local-level planning in integrated rural development, divergent views were expressed about that specific level at which such an exercise ought to be undertaken. In some countries, the most convenient level was identified as the district, as the focal point to aggregate the needs and converge the participation of the rural populace. In other countries, a higher or lower level relative to the district was considered most appropriate. The Expert Group, however, was clearly in agreement that selection of the most appropriate level would have to be determined by each country taking into consideration the particular conditions prevailing. However, some considerations that could objectively determine that process would include:

(a) Its possibility to facilitate direct and indirect participation of local people in the planned development process;

(b) Its coverage of an area viable enough to allow the provision of essential services and upgrading the managerial and productive skills of the rural poor through group action to assist in the mobilization of resources to activate a locally-inspired development programme;

(c) Its coincidence with the lowest level of the administrative hierarchy.

While the selection of the appropriate level at which local planning was to be formulated would, in the first instance, have to be influenced by pragmatic considerations, the Expert Group was mindful that if the local-level planning exercise was aimed at the rural people and their involvement, the basic environment most appropriate would be the village. It was at that level, moreover, that heterogeneous groups and strata could be most meaningfully taken into consideration in order to provide for plural participative links in the processes of local-level planning.

One related aspect that drew the attention of the participants was the definition of the objectives of local-level planning as that exercise was but a means to attaining some goal or goals. Some experts felt that the focus of local-level plans should invariably be income-generating activities ; others expressed the opinion that the appropriate development of social services was as much a legitimate concern of local-level planning. Yet another group felt that the specific goals of local-level planning should be coterminous with the goals of rural development as, in the final analysis, the evident desire was to promote integrated rural development with the active participation of the rural poor. Thus, the enhancement of the quality of rural life should be the legitimate goal of local-level planning and, for the attainment of that goal, the rural poor should be given the opportunity of participating in decision-making. In that connexion, the schedule of goals of local-level planning identified in the report of the UNAPDI Consultative Meeting on Alternative Strategies for Development with Focus on Local-level Planning and Development had a bearing on the topic.

Functions and Content of Local-Level Planning

Given that the local-level plan formulation could not be undertaken in a vacuum, it would appear that the process should integrate into it the means whereby adjustments and modifications could be made to national, regional and sectoral planning guidelines while working within the framework of available resources. That would ultimately lead to a developmental scenario at the local level consistent with the specific needs of the people, growth potential of the area and budgetary allocations available. While that was a generally acceptable proposition, some strong feelings were expressed that even such a formulation would presume upon and compromise the integrity of the decision-making prerogatives of the rural people. It was, however, counter-proposed that, in practical terms, some element of compromise would appear inevitable as certain development projects by their very nature would transcend the needs of individual villages or districts. Furthermore, national planners would have to formulate broad guidelines not only of planning strategies consistent with prevailing political philosophies but also of equitable allocations of resources across the country.

The Expert Group took cognizance of the Working Group on Block-level Planning in India which had suggested that some of the broad components of local-level plan would include :

(a) Production programmes for the target groups in the area in the light of resource potential ;

(b) Manpower planning and skill development in relation to the production programme ;

(c) A programme for the provision of basic minimum needs ;

(d) A programme for institutional support.

The Expert Group was of the view that as far as possible those programmes should be managed by people's organizations.

Some experiences of the countries represented threw useful light on the functions and content of local-level plans. In the *Markaz* model in Pakistan, for instance, the first step in the planning process was a diagnostic survey elaborating secondary data and supported by an extensive process of consultation with villagers. Dialogues of that type were generally instrumental in providing perceptions of problems and requirements in a different light and could lead to the provision of projects that would have been overlooked by more distant planning processes. To assist in the planning and implementation process, the organization of local institutions was of the highest priority, as was the location of an administrative centre at the *Markaz* level to provide villages with ready access to inputs, services and training.

The design of local participation through institutions was stressed by several experts who identified the danger that such institutions could readily be subverted from their planned goals by becoming mechanisms for allocating patronage or monopolizing inputs and services provided by the Government, or be dominated by the local elite. It was also felt that, in dealing with the existing government strutures at the local level, there was a need to forge greater interagency co-ordination. Most experts agreed, moreover, that a socio-economic profile or diagnostic survey would effectively aid in articulating the needs of the rural poor.

General consensus was apparent that local-level planning would, in practice, imply the location of a development centre, including people's organizations and development departments at that level, to provide co-ordination, to mobilize and organize people's participation, to plan for infrastructure, and economic and social development, to facilitate the undertaking of a diagnostic survey or village profile and to supply training inputs to village-level cadres.

Effective Participation of the Rural Poor

One of the crucial linkages in local-level planning was the

participation of the rural people in the processes of integrated rural development. Experience had, however, shown that the process of eliciting participation through local institutions could severely compromise, *bona fide* participation by the rural majority because of hierarchical or sectional interests and groups. The need for the formation of groups for small farmers and landless labourers was also manifest and attested to by the experts. How, then, was the process of organization and institutionalization of the rural poor to be programmed ?

Some experts felt that such institutions should not be bureaucratically inspired, as past experience had amply indicated the dangers of creating a dependency of the poor on the Government. Others expressed the view that non-Government organizations should rightly be accorded the role of initiating the organization of the rural poor although the dangers of impinging upon the sensitivities of Governments should not be overlooked. The experts felt that perhaps the only feasible way to establish links to the rural people country-wide was, to start with, through the existing bureaucracies.

If that was to be the strategy of organizing and institutionalizing people's participation in local-level planning, certain dangers were inherent and should not be lightly disregarded. These included the possibility that even where the political will was manifest in desiring a devolution of the planning prerogatives to lower levels, the bureaucracy could subvert that desire. Further, civil bureaucracies could be self-serving or not highly motivated in wanting to devolve authority to the local people as that militated against the traditional civil bureaucracy's ethic of control and regulation. In such a situation, it was considered that a possible rational solution in reversing the historical patron-client mentality would be to :

(a) Introduce performance criteria for bureaucrats related to predetermined goals ;

(b) Provide incentives to bureaucrats in the form of better promotional prospects and economic compensation ;

(c) Enhance the accountability of the bureaucrats to the rural people they were serving.

A case in point was indicated by the situation in Nepal where current thinking related to the introduction of incentives and performance criteria for bureaucrats to serve in the rural districts and make them accountable to village and district *Panchayats*. A similar process had taken place in Sri Lanka, where the district minister system had been introduced as a clear manifestation of the devolution of political authority to the district.

In so far as the rationale for eliciting the support of the rural people for rural institutions was concerned, clear incentives should be made available. The success of those efforts might depend on the following factors :

(a) Projects undertaken should be economically viable and provide tangible gains in the short run ;

(b) While members might be called upon to make sacrifices (in the form of labour or capital investments), the net benefits should clearly exceed their sacrifices and the benefits of operating individually.

EXAMINATION OF MAJOR CONSTRAINTS ON LOCAL-LEVEL PLANNING

The specific constraints preventing the full operation of the process of local-level planning for integrated rural development were identified as follows :

(a) If national policies to decentralize the processes of rural development were not undertaken with political will, that could severely limit the achievements of local-level planning, for instance in the form of budgetary difficulties, land-ownership structures that facilitated exploitation, unfavourable terms of trade leading to low purchasing power, and lack of coverage of risks involved owing to uncertainties in agriculture resulting in less than optimum production;

(b) The civil bureaucracies largely operated within law-and-order and control parameters rather than the promotion of rural development or the sharing of authority and responsibilities with the rural poor. Thus, even with the best of intentions, bureaucracies with traditional orientations could hamper the processes of local-level planning;

(c) The lack of optimum methods of organizing and institutionalizing the participation of the rural people, particularly where rural communities were heterogeneous in nature and structure;

(d) The limited managerial skills of local people and organizations;

(e) The inadequate planning and management competencies of bureaucrats operating at the local level;

(f) The difficulty of matching local needs with nationally determined priorities and of relating that to available resources and budgets;

(g) The long-established dependent attitudes of the rural population.

SUGGESTIONS FOR A PROGRAMME OF ACTION

In the light of the experiences of the countries represented at the Expert Group Meeting—taking into account the acute need to involve the rural people, particularly the rural poor through their own organizations in local development, and recognizing the constraints facing the full operationalization of the processes of local-level planning for integrated rural development—the Group identified the following areas for the attention of Governments and of the international community.

Training

In several of the countries in the region, there was an evident need to improve the planning capabilities of bureaucrats as much to assist in the over-all process of national and regional planning as to provide planning inputs in the local-level planning exercises. Given the varying environments obtaining in countries in the region, training programmes would have to be country-specific in nature.

A greater need was to improve the capacities of government functionaries and the rural people to plan for local-level integrated rural development. Such training programmes which could most effectively be organized for trainers first, for eventual replication throughout a country, would have to give specific recognition not only to the formulation of local-level plans but also to the need for evaluation techniques and mechanisms to measure the impact of local-level planning and the contribution of the people in the planning process. Such training programmes, when eventually undertaken for local-level functionaries, should necessarily be conducted in the language or dialect used in a particular locality.

Training programmes were also required to improve the management capacities of local functionaries in order to develop their abilities beyond their traditional responsibilities. Furthermore, training programmes to enhance the managerial skills of leaders and other officials of rural organizations had to be successful economically if they are to exert an impact upon local-level planning exercises.

In the above training programmes due attention should be given to the role of women in rural development.

Research and Studies

To strengthen and complement ongoing research and action programmes in different countries and institutions, further research programmes, including action research, were required to understand the complexities of the processes of local-level planning and assist in the development of training material. Some critical areas and linkages which required investigation and clarification were :

(a) Strategies required to mobilize, organize and institutionalize the participation of local groups in the planning exercise;

(b) Ways in which existing government machinery could be modified or improved in order to facilitate the processes of local-level planning;

(c) Methods by which planning functions could be effectively and expeditiously decentralized to make local-level planning a reality;

(d) Procedures by which budgets could be operationally decentralized to local levels;

(e) Ways by which incentive systems and performance criteria could be formulated and implemented in order to induce government functionaries to work in rural areas, devolve decision-making programmes to the rural people and assist in the organization of local institutions;

(f) Improvement of the methodology of the planned development process, with special attention to :

(i) Simplified planning techniques;

(ii) Improved techniques for diagnostic and community self-surveys;

(iii) Development of simplified systems of monitoring and evaluating rural projects with the participation of local functionaries and the local people;

(iv) Improving the consistency of the content of local-level plans in relation to the objectives;

(v) Development of local-level statistics and their integration into regional and national strategies;

(g) In some cases, assistance might have to be rendered in the strengthening capability of countries concerned in the area of research and evaluation related to rural development problems;

(h) Action research projects (i) to understand and analyse the implications of uncertainties and risks in agricultural and allied activities on the motivation of small farmers in organizing themselves in groups, and (ii) to understand the likely impact of risk coverage on the motivation and production enhancing activities of small farmers;

(i) Action research projects to develop appropriate institutions and mechanisms such as "labour certificates", for the mobilization and more intensive use of surplus labour available in rural communities for local-level development, with special attention to landless labourers;

(j) Implications of local-level planning and development on higher levels of planning, or linkages between local-level planning and other levels of planning, from the perspective of local-level development.

Exchange of Information

The creation of a clearing house for information, documentation and experiments on local-level planning in the region seemed timely, especially in the light of the experiences of several countries in the region. Such a clearing house would collect and disseminate relevant government and non-government documentation relating to local-level planning experiences and case studies. It was suggested that greater co-ordination could be established among United Nations agencies for cataloguing existing documentation of country-specific experiences related to local-level planning.

A thorough assessment of successful experiments in local-level planning to indicate the procedures followed, results achieved, obstacles faced, etc., could generate useful insights and guidelines for those persons interested in adopting such an approach in their respective countries.

Study Tours, Seminars and Workshops

To complement the preceding subsection, study tours could be arranged to afford policy-makers, planners, local-level functionaries and others the opportunity to study at first hand experiments currently being undertaken in the region. They could lead to seminars and workshops where the specifics of local-level planning processes could be discussed by those involved in such processes in the context of integrated rural development. Furthermore, the problems and possibilities of integrating the bottom-up approach of local-level planning with the broader processes of national or regional planning could be explored.

Conducting Experiments in Local-Level Planning

Where Governments were receptive to the idea, field experiments of local-level planning could be conducted in selected countries as a basis for testing alternative approaches to local-level planning and developing appropriate methodologies for that process. That could be useful in broadening the experiences of those involved in local-level planning exercises and contribute to the development of training material and training programmes for local-level functionaries.

ADOPTION OF THE REPORT

The report was adopted on 10 November 1978.

REFERENCE

1. Consultative Meeting on Alternative Strategies for Development with Focus on Local-Level Planning and Development, Bangkok, 31 October—4 November 1978.

NINE

ORGANIZATION AND CONTENT OF LOCAL-LEVEL PLANS

B.W.M. Van Dusseldorp

Introduction

LOCAL-LEVEL PLANNING is propagated as an important tool in realizing integrated rural development with emphasis on the weaker groups in society. But how local-level plans have to be organized and what the content should be is not always clear. This paper makes an attempt to give a tentative idea of the possible organization and content of local-level plans.

First, the major functions and activities of local-level planning are indicated. Then the possible content is discussed and finally the consequences for local-level planning units are indicated if local-level plans are made in the way proposed.

Some of the points discussed hereafter are an elaboration of issues mentioned in the note by the secretariat on local-level planning for the Expert Group Meeting.

PLACE, FUNCTION AND ACTIVITIES OF LOCAL-LEVEL PLANNING

Local-level planning is the first level of integrative planning that takes place above village, project and detailed programme planning.

The major functions of level planning are linking and integration :

(a) To adjust general guidelines from higher levels in such a way that a development programme will emerge that is specifically for the area for which the level planning is made (top-down linking);

(b) To adjust programmes for lower levels in such a way

that they meet the goals, objectives and resources indicated in the higher level plans. This adjustment must be done so that the lower-level proposals are protected as much as possible (bottom-up linking);

(c) To integrate and co-ordinate the multitude of lower-level (village, project and detailed programme) proposals into one consistent local-level plan that is meeting the goals, objectives and targets from higher as well as lower levels, that can be implemented with the available resources and indicates the geographical location of the development activities (horizontal linking).

It is assumed that local-level planning must avoid duplication with other planning activities and should limit itself to its linking and integrative function. It should whenever possible avoid getting involved in project and detailed programme planning. It can, however, when necessary, identify new projects and detailed programmes that are needed in order to fill gaps with present or future patterns of action, but leave elaboration to the relevant agencies.

If the major functions of local-level planning are understood as mentioned above, its main activities are the following :

(a) To collect all the plans under implementation or preparation by all agencies (government, semi-government, voluntary, professional or private) that are involved in planned development activities of relevance for the area;

(b) To collect all goals, objectives and targets indicated in higher and lower level plans;

(c) To obtain information on the resources available in and allocated to the area;

(d) To indicate gaps with the existing or proposed pattern of development activities and identify new projects and programmes that have to be elaborated by the relevant agencies in order to reach the goals and objectives;

(e) To obtain information on the results of previously planned development activities in the area and the reason for their success or failure;

(f) To integrate the above information into:

(i) an overall analysis of the present situation;

(ii) a general description of the desired future situation on the basis of the accepted goals, objectives and targets;

(iii) a coherent set of programmes of action that can be implemented with the available resources and lead to the desired future situation.

ORGANIZATION AND CONTENT OF LOCAL-LEVEL PLANNING

Relevant methodological aspects are discussed in more detail in the following sections dealing with analysis, target setting, plan formulation and evaluation.

Integration and Linking in the Analysis of the Current Situation

Integration and linking should find expression in the description and analysis of the current situation and the assessment of major potentials and constraints for local development. Not only is an over-all picture of the current situation required but also the processes that brought this specific situation into being have to be indicated. In local-level planning special attention has to be given to major linkages between these processes.

In the analysis of the present situation, a succinct description of the existing physical, economic and social structure has to be given. This structural analysis gives a convenient vehicle, for integration of data (see also table below). Since local-level planning is aiming at greater equity, the focus of the analysis has to be on the interrelated processes that have created the present inequity and are maintaining it.

For a good analysis of the present situation, a considerable amount of data is necessary. In the collection of data, local-level planning units should rely heavily on the officers from those departments that are involved in development activities in the area. By involving them in the provision of data, often via the projects and programmes prepared by them, a process of integration is started that is of importance for later stages in the planned development process, e. g. the planning and implementation of the local-level plan. Only when there are major gaps in the data already collected should the planning unit do investigations itself or, preferably, assign surveys to professional institutes or government agencies.

The main function of level planning units in making an analysis of the present situation is not to collect data, but to indicate existing gaps in the available data, to check the data received by others and to integrate these data into a dynamic analysis of the present situation. Checking data should not be seen as desk research or statistical checking only; it requires also intensive trips into the field.

Integration and Linking in the Description of the Desired Future Situation

Another important function of level planning is to integrate the goals, objectives and targets given by higher, lower level and sector plans into a consistent set of goals, objectives and targets for the area.

However, goals, objectives and targets give only a segmented view of the desired future situation. Thev have therefore to be translated into a structural description of the desired future situation. As it is impossible to describe the future situation in one over-all structure, it is necessary to indicate the main structure and sub-structures (see table below). These structures, superimposed on each other, are supposed to give a good idea of the future situation that will emerge when the goals, objectives and targets are realized. Special attention must be given to the physical structure. These structures and substructures are closely linked to each other and some structures have to be realized before others can come into existence.

The question arises, when it comes to level plans aiming at equity, if it is wise to indicate explicitly the future socio-economic structure. It can arouse the anxiety of powerful groups and lead to resource mobilization that will thwart the major goal of such plans. In other words how honest can the planner be without endangering the objectives of the plan?

Structures and substructures to be described in local-level plans (in the analysis of the current and description of the desired future situation)

Economic structure	*Main elements* : the output and employment in each sector; income distribution.

Socio-economic structure	Socio-economic structure	*Main elements*: type of production organizations; large factories, plantations, small-scale family-owned industries or farms, production co-operatives, or a certain combination of these production units. An indication of the number of people involved in the various production organizations and their output.
	Social structure	*Main elements*: the various groups, either based on consanguineous or territorial criteria, and their inter-relationship, leadership, and power structure. Rough indication of the number of persons involved in the various groups of the social stratification.
Administrative structure	Administrative and participation structure	*Main elements*: structure of the governmental organizations, their interrelationship, number of people involved and their output. Organization of participation, village councils, district councils; their major task and functions.
	Structure of the health sector	*Main elements*: types of service units, their inter-relationship, people employed in these various units and their output.
	Structure of the educational sector	*Main elements*: types of schools, their interrelationship, teachers employed and the school population in the various school types and their output.

Physical structure	Land use pattern	*Main elements* : what areas of the region are used for various types of agriculture, for forestry, fishing, animal husbandry, recreation and conservation, mining, industry, housing, etc.
	Pattern of service centres (including urban)	*Main elements* : the types of service centres (central places) and their interrelationships and threshold values. The services (such as schools, clinics, hospitals, shops, banks, offices for the administration, extension services, markets, industrial sites, etc.) belonging to the various types of centres. Rough indication of population living in, and served by the centres.
	Transportation and public facilities	*Main elements* : road and railroad pattern, harbours, airfields and their location. electricity and drinking-water facilities, radio and telephone facilities.

Integration and Linking During Plan Formulation

By comparing the existing and the desired future social, economic and physical structure, it becomes possible to indicate the programmes of action necessary in order to travel from the present to the desired future situation.

However, it is not the task of the local-level planning unit to draw up all these programmes of action by itself. Here also the planning unit should rely on officers working in departments on development activities in the area. The planning unit has to make an inventory of the ongoing and planned programmes of action. After analysing this information, the local-level planning unit has to indicate gaps in the existing pattern of planned development, to fill the gaps either by directly preparing relevant programmes of action or, preferably, by assigning this task to other agencies.

The tasks of the local-level planning unit are :

(a) To match the resources available with the demand for resources arising from the numerous projects and programmes proposed by various agencies and the additional proposals that the local-level planning unit may consider necessary to fill certain gaps. Here the unit has a dual task and loyalty is involved. On the one hand, it should not hesitate to advocate a greater resource allocation to the area when it finds new potentials that are worth developing and that were not known before by higher level planning units. On the other hand, it must be realistic and consider that funds from higher levels are limited. An important task would be to explore the possibility of mobilizing local resources to realize those sound projects that could not be financed by external funds. It is of great importance that the local-level planning unit does the matching of available and required resources, in close co-operation with the relevant departments, who will actually be responsible for implementing the projects and programmes included in the local-level plan;

(b) The second task is to indicate the linkages between the various programmes. For instance, increasing agricultural production requires a credit and supply system for the new inputs as well as a better road system, storage and marketing facilities. The educational system must train the people with capabilities required for the future manpower structure. Social and economic facilities should be brought together in rural service centres and the latter should be related to the transport and electricity networks.

To what extent should the local-level plan be integrated? Should it consist only of a list of projects matching the funds available without any effort to integrate them? Or should only the main relations between the major development activities be indicated? If this is the case, one may wonder what local-level is all about and whether it is worth the effort. On the other hand, a programme of action translated into an integrated and computerized network programme would not only be too costly but could not be managed by most agencies.

The local-level plan must assess :

(a) The importance of linkages between the various projects and programmes for implementation of the plan;

(b) The willingness and capacity of the various agencies involved in the planning and implementation of development activities in the area, to co-operate and co-ordinate their activities with one another;

(c) The co-ordination and implementation capacity of the local-level administration. On the basis of this information the local-level plan must indicate what has to be co-ordinated and how the co-ordination of the implementation should be organized.

When local-level plans are made in order to let the weaker groups be the first to profit from the expected benefits of the plan, integration and linking of the various development activities become even more important.

The rural poor have a low risk absorption capacity, especially over an extended period. Even small delays in providing inputs or transport, and a market system that cannot readily absorb at reasonable prices their increased production, can mean disaster to them.

Projects aimed at the rural poor are fragile projects that require considerable support. Unless major and minor constraints in the physical, economic, administrative and social infrastructure are removed opportunely these fragile projects will most likely fail.

Local-level planning, provided it pays attention to the most relevant linkages and uses these in the design for an integrated programme of action, can be a useful instrument to get an integrated rural development programme going for the benefit of the rural poor.

Integration and Linking During Implementation

The implementation of local-level plans has two aspects :

(a) The detailed programming and implementation of all activities included in the plan;

(b) The co-ordinated implementation of such activities.

The first implementation activity is the responsibility of the lower echelons of departments and other relevant agencies. The second is in most cases the responsibility of the head of the local level administration (e.g. the District Officer). In practice he needs assistance to perform this function properly. This assistance, which can be provided by a separate co-ordination unit, or by the local-level planning unit has to be assured. If the second solution is chosen care must be taken that the planning unit is not so much involved in the day-to-day activities of co-ordination that it is no longer in a position to pay attention to the adjustment of the ongoing level plan and the preparation of new level plans. If the planning unit is charged with this task it should have a co-ordination officer.

A certain co-ordination capability is an essential prerequisite for local-level planning. The level of integration indicated in local-level plans should be adjusted to the co-ordination capability and power of the administration. Planning should not start without a guarantee that at least a minimum of effective co-ordination capacity is available or could be made available. Otherwise, local-level planning will be a waste of money.

Evaluation of Local-Level Plans

The local-level planning unit has to be involved in evaluation, at least for the internal interim and final evaluation of its own performance. The question can be raised whether this unit should also be appointed as the evaluation agency for the local-level implementation activities. In theory an independent unit would be preferable. Taking into account the scarcity of manpower and the importance of free flow of information from the evaluation unit to the planning unit, a combination could be considered. However, there remain certain disadvantages, such as the planning unit evaluating its own work; the working load that will arise from this extra activity, that can endanger its planning work. If the planning unit is also to be charged with evaluation it should have an evaluation officer who is not too much involved in planning activities.

SOME CONSEQUENCES FOR THE PLANNING MACHINERY AND ITS ENVIRONMENT

If local-level planning is understood as described above, this has consequences for the conditions under which it can work effectively :

(a) The quality of the local-level plan depends mainly on the quality of the projects and programmes it is integrating. This means that high priority should be given, if necessary, to improvement of the planning capacities of the lower echelons of the various departments;

(b) As the integrated plans and programmes of a local-level plan have to be prepared and implemented by other departments, it is great importance that these departments, via their planning/implementation officers, are a part of the local-level planning unit;

(c) Local-level planning units can be organized in several ways. For each area where local-level planning takes place, a planner may be assigned to prepare the plan, in co-operation with the various departments and, if necessary, with experts from a higher level on an *ad hoc* basis;

(d) Local-level planning units can be set up consisting of a small core of professional planners and officers of relevant departments assigned to it on a part-time basis. As such units in most cases have to serve several areas they should be situated at a higher level in the administration (e.g. if the local-level plans cover subdistricts, the planning unit could be located at the district office);

(e) Other agencies involved in the development of the area (private, professional, voluntary) should be involved in the preparation of local-level plans;

(f) As previously mentioned, local-level planning can be of little avail if it is not matched with a co-ordination capacity for implementing integrated programmes at the local level.

TEN

THE STATE OF THE ART IN SELECTED COUNTRIES OF THE ESCAP REGION

THE INTEREST in local-level planning for rural development is growing rapidly in the countries of this region, but there is still a wide gap between what the experts have recommended and what is actually practised (as reflected in the following country papers).

In essence, what the experts recommended could be summarized as follows. If the rural masses, which often constitute a majority of the population, are to be effectively involved in the development process, a decentralized approach is required that will take full advantage of the human and material resources available in the rural areas. This approach must obviously be supported by appropriate national strategies; but local-level planning is necessary to achieve a coherent pattern of rural development on a sustained basis, aimed more especially at improving the productivity and living conditions of the poor people.

It is generally agreed that popular participation will be best achieved at the village level; and within each village it may be necessary to develop special interest groups (e.g. among small farmers or landless labourers) if these groups are to be given a fair opportunity of participating in development programmes.

This micro approach must, however, be reconciled with the need to develop the required local services over a broader administrative area (including procurement of inputs, credit, marketing, extension of agriculture, training of local leaders, primary health care, education facilities and so on). In several countries of the region, the "district" is considered the most appropriate area for planning and implementing a coherent pattern of rural development programmes; in other countries, the area selected for such purposes is the "subdistrict" covering a population of about 100,000 (e.g., the *block* in India, *Markaz* in Pakistan, *Kecamatan* in Indonesia).

Local-Level Planning in Selected Countries

For the purpose of the Expert Group Meeting, a distinction was made between (a) national planning; (b) intermediate (or regional) planning; (c) local-level (district or subdistrict) planning; and (d) village programming. It may be useful to summarize at this point, the current practice in the eight countries for which country papers have been prepared.

India

For the purpose of rural development, India has been subdivided into 5,004 blocks (subdistricts), and more recently a major programme has been launched for developing "block-level plans" in 2,000 of these blocks within the next five years. It is felt, however, that the planning team for "block-level planning" could be more realistically located at the district level.

Indonesia

In the past, the district (*Kebupaten*) was considered the main unit of decentralized administration in the rural areas; but more recently the new policy has been to promote the *Kecamatan* (subdistrict) as the operational basis for integrated rural development. At this level, the local development working unit (UDKP) is responsible for setting up priorities, integrating the sectoral programmes and co-ordinating their implementation, under the leadership of the subdistrict officer (*Camat*).

Malaysia

In recent years the Government of Malaysia has been very much concerned with rural poverty and regional disparities. Strategy has been pursued largely on a sectoral basis (e.g. rubber, rice etc.); and some regional plans have been prepared for less developed areas. At the district level, the District Officer has a responsibility for ensuring the implementation of rural development programmes; but these usually originate from various departments and authorities and do not necessarily form a coherent district plan.

Nepal

This country is subdivided into 75 districts (average population : 200,000), which play a pivotal role in rural development. Since 1960, a *Panchayat* system has been instituted at the village and district levels for purposes of local administration and development. Recent experiments have tested the possibility of organizing small farmers at the village level and of using subcentres (within each district) for decentralizing project activities.

Pakistan

For the purpose of rural development, the country has been subdivided into 600 subdistricts (*Markaz*) including about 70 villages and an average population of about 100,000. Multipurpose farmers' associations are organized at the village level and services and training facilities are provided at the *Markaz* centre.

Philippines

In recent years, the accent has been put on regional development (in 13 regions), and also on the preparation of integrated rural development projects (supported by external donors) in some "priority" areas that were less developed but offered a good potential for rapid development. In other areas the initiative for local development has been left to the initiative of sectoral departments (e.g. agriculture, irrigation, agrarian reform, local government).

Sri Lanka

This country is subdivided into 22 districts with an average population of 600,000. A district minister is to be appointed in each district, which will enjoy a large degree of autonomy. A development plan will be formulated for each district. Some experiments are concurrently made to revive the initiative of 1971 for developing integrated rural programmes at the subdistrict level (Division/Electorate) in closer contact with the villages.

Thailand

The country is divided into 71 provinces (*Changwat*) and 540 districts (*Amphoe*) which are, in turn, subdivided into 4,500 communes (*Tambon*) and 50,000 villages (*Muban*). So far, programmes for rural development have been formulated more or less on a sectoral basis. Some attempts have been made at regional planning in the north, north-east and south. In the context of the Fourth Plan (1977-1981) a mandate has been given to formulate provincial development plans in all of the 71 provinces (with an average population of about 600,000).

The Challenge Ahead

The above countries have accumulated a broad experience in rural development over the past two decades or so. In most of these countries the production of food and agriculture has increased significantly, but it is not evident that the conditions of the rural masses, especially the small farmers, sharecroppers and landless labourers, have been significantly improved as a result. Most Governments are officially committed to do something about it ; but so far these objectives do not appear to have been translated into effective enough strategies at the grass-roots level.

During the coming decade, local-level planning (at the district or subdistrict level) could play a useful role in crystallizing such strategies, provided that the process is strongly supported by appropriate national policies (including resources allocations), that the local people (especially the rural poor) are well organized to participate in the development process and that the local administrative machinery is fully committed to assist in the implementation of the rural development programme in partnership with the local people. Otherwise, the local plan might turn out to be little more than an academic exercise.

Assuming that the above conditions will gradually materialize, ESCAP, in co-operation with the other agencies concerned, intends to provide systematic support to member countries wishing to improve their capacity for local-level planning in rural areas. A first step will consist in preparing detailed models of local-level planning already operationalized in different settings. This will be followed by study tours to actual experimental projects of local-level planning in selected countries of the region. Other contributions will be to assist in testing new experiments in local-level planning in interested countries, to assist in training local level planners, and generally facilitate the exchange of information and experience in this field among the countries of the region.

ELEVEN

INDIA

K.P.A. Menon

Past Experience : A Summary*

INDIA has had broad experience of the local approach to rural development since the community development programme was launched in 1952. During the period of the Second Five-Year Plan (1956-1961) an effort was made not only to decentralize the administration at the district and block levels but also to promote the establishment of a three-tier system of the village, block and district *Panchayats* (statutory, elective local bodies).

For a while, the accent was put on Intensive Agricultural Area Programmes meant to increase food production through the introduction of high-yielding varieties of crops. The benefit arising out of these programmes generally accrued to the affluent farmers and better endowed areas. Therefore, some special programmes had to be launched to help more directly the small farmers and drought prone areas. Other programmes were meant for the development of irrigation projects (Command Area Development). All these programmes led to some degree of "local-level" planning, although the emphasis tended to shift depending on the type of programme (e.g. agricultural production, irrigation system).

During this period, concrete steps were taken to remove regional imbalances and to extend the benefits of development programmes to the weaker sections of the society. In 1969, the Planning Commission issued a set of guidelines for the formulation of district plans. Such plans were effectively prepared by a number of state governments. More recently, in 1977, new impetus has been given to area planning in order to intensify rural development and promote fuller employment as a matter

* Summary prepared by ESCAP Secretariat.

of high priority. Two thousand blocks out of a total of 5,004 blocks in the country have been selected for a programme of intensive development based on block-level planning over a period of five years (1978-83).

Current Trends

The issue of area planning at different levels came into sharp focus following the decision of the Government to have an intensive rural development programme aimed at full employment within a definite time frame of 10 years. Addressing the Seminar on Block Level Planning for Full Employment held at New Delhi on 17 and 18 August 1977, Professor D.T. Lakdawala the then Deputy Chairman of the Planning Commission, mentioned that area planning had the great advantage of compelling the planners to look more closely and concretely at the problems at local levels and to devise measures for the optimum utilization of the natural and manpower resources to suit local potentials, needs and problems. Addressing the same seminar, the then Prime Minister of India, Mr. Morarji Desai, referred to the inadequate attention given to rural development and deteriorating problems like disruption of life in the countryside and the exodus of the rural population to the towns, which resulted in the creation of slums. To make up for this neglect of rural development, it was necessary to have a reorientation of efforts in order to catch up with the problems and to ensure faster progress in the rural areas. Talking about block-level planning, he mentioned that plans unrelated to resources available had no relevance. Block-level plans must be essentially practicable so that they could be realized, acceptable to the people for whom they were intended and related to the resources available in the country. He also felt that the non-official agencies would have to be involved in the whole process since they could make the average villager interested in the programme and take to it enthusiastically and willingly. This was a task which could not be done by government agencies effectively. He also felt that village *Panchayats* should be brought into existence in all the States and at the same time steps taken to ensure that political factors did not dominate them. He also emphasized the importance of decentralization in the whole process of planning and the need for adequate and appropriate delegation of functions and powers right from the centre at the top, through the State, down to the village *Panchayats*.

That seminar was only the beginning of a chain of events all of which had a great bearing on local-level planning. By a government resolution dated 12 December 1977, a high-level committee was appointed to inquire into the working of the *Panchayati Raj* institutions and to suggest measures to strengthen them so as to enable a decentralized system of planning and development to be effective. This committee was headed by Asoka Mehta, a former Deputy Chairman of the Planning Commission. Earlier, on 15 November 1977, the Planning Commission had appointed a working group under the chairmanship of Professor M. L. Dantwala, an eminent economist, to prepare guidelines for block-level planning in the country with a view to achieving full employment during the subsequent 10 years. Another committee, headed by B. Sivaraman, member of the Planning Commission, was simultaneously studying the role of voluntary organizations in the implementation of block-level plans and programmes of rural development.

Recommendations of the Asoka Mehta Committee

The committee has pointed out that democratic institutions must articulate their structures in terms of their functions which, in their turn, were determined by the possibilities of development. Over the past 20 years, the horizon of development had greatly widened in the wake of numerous changes in the various facets of the economy. Therefore, the committee felt that the institutional, structural and functional contours of *Panchayati Raj* had to be in conformity not only with the accelerating pace of development but also with the developmental strategies and policies that had evolved over a period of time. The *Panchayati Raj* institutions had to be so designed as to become effective channels of people's participation in the process of growth. With that basic premise, the committee devoted a whole chapter to the "Dynamics of development".

After detailed discussions on the dynamics of development the committee dwelt upon the need for technical assistance and organization of a high order that was necessary to sustain the momentum. While the State government had the requisite competence in that regard, a process of administrative decentralization had already started at the district level. Competent experts of different departments were already stationed at the district level and their capabilities were also being upgraded in both the industry and agriculture, sectors, In agriculture, subject-matter specialists were being stationed at the district headquarters. The District Industries Centres that were being set up would have a component of experts with different specializations. Setting up of a district planning unit with technical expertise in several fields was also under consideration. The district was also traditionally a unit of administration to which all were acclimatized. Further, it was the key planning unit for evolving over-all strategies, resource allocation, including credit, and delineation of different types of projects. The committee had, therefore, recommended the establishment of a strong *Panchayati* tier at the district level to respond both to organizational needs with in the district and to the demands of emerging linkages with the state.

Below the district level, the committee felt that the balance between technological requirements and the possibility for meaningful participation by the people in development management could best be achieved by grouping a number of villages to constitute the *Mandal Panchayats*. That could ensure economic viability and enable the people's representatives to exercise democratic supervision over the large number of micro-projects which were to be implemented at the local levels in the coming years. Those projects were not entirely village based, but would, in the opinion of the committee, require a span of technology beyond the capacity of a village *Panchayat*. The attention to the family-based programmes, often covering as many as 400 families, needed larger units to tackle than were available under the system of village *Panchayat*s. The committee visualized that such *Mandal Panchayats* would normally cover a population of 15,000 to 20,000 each and would also facilitate the forging of necessary linkages with schemes of development of focal points and growth centres ensuring efficient management of the rural-urban linkages.

The committee did not favour a separate local administration at the block level or village level. Taking note of the fact that block-level *Panchayat Samitis* already exist in a number of States, the committee recommended that those could be converted into non-statutory committees of the *Zilla Parishads*. When *Mandal Panchayats* were organized and became active, most of the functions of the *Panchayat Samitis* could be taken over by them. At the village level, the people would be involved in *Mandal Panchayats* through village committees which would look after municipal functions and related welfare activities.

As regards the process of planning, the committee concluded that there was considerable scope for decision-making at the district level. The level was broad enough to consider policy options and immediately below the State that was a strategy level for economic planning. The committee did not however, make detailed recommendations regarding the structure that was necessary at the district level for the preparation of district plans primarily because that had been dealt with in detail by the Dantwala group. The block level *Panchayat* would, so long as it continued, play an important role in supervising the execution of development programmes. The committee viewed the problem from the functions that could be appropriately discharged by the *Zilla Parishad*, the *Panchayat Samiti* and the *Mandal Panchayat*.

According to the committee, the structures of the *Panchayati Raj* institutions provided a suitable machinery for the effective functioning of multi-level planning. Progress had already been achieved in institutionalization and standardization of techniques and procedures for district planning. When the district was made the key unit of decentralization, the task would also become easy. Guidelines were likely to be issued for block-level planning very soon. The committee felt that that was a suitable beginning and the process itself would develop district planning It had particularly been emphasized that emphasis on planning at the district level was not incongruous with the current emphasis on blocks. Each was a flow into the other. A format, strategy, resourcc, planning, total infrastructural problems and the necessary accent on different agro-climatic zones within a district, etc., would have to be covered in district planning. The production and employment programmes should be so designed at the block level as to fit into the totality of the picture. The other interblock ramifications would also be taken into the purview of the district plan. Where a block itself could not cover small urban areas, the district plan would. The *Zilla Parishad* would also provide the much needed political corrective to the techno-economic plans. It is interesting to mention in that connexion that the Dantwala group on block-level planning also emphasized the point that district plans and block plans were to be supplementary and complementary to each other and not competitive.

The state government would have to set up a machinery for examining each district plan in order to formulate the State plan. It would be necessary to have a detailed dialogue with the districts, represented by the *Zilla Parishad* and the planning group. The total State plan structure would evolve only after those discussions. Since the territorial plan of the existing district would include a number of densely populated urban centres, the district planning machinery would have to be specially geared to take care of the urban-rural continuum so that it could plan for the regular interchanges of goods, services and human population between the towns and the villages. In the case of districts which contained a large number of urban centres, those rural-urban exchanges would be sufficiently large to demand special treatment. In the case of small towns servicing a larger number of villages, the cluster level *Mandal Panchayats* would have to gain interlinkage with the economic and developmental aspects of the urban bodies.

The committee also felt that the block-level planning exercises would really help fill in the format of the district. The production and employment plans could be sharpened at levels lower than the district. To begin with, the role of the *Mandal Panchayat* would essentially be one of promotion and implementation. It was not expected that they would be able to develop viable plans of their own to this stage. With the assistance of the district planning cell, the concerned departmental functionaries stationed at the *Mandal* level would be able to build up suitable projects. The *Mandal Panchayats* would have to supply the necessary information about the various schemes and also suggest the potential areas of development and the type of requirements that had to be met. That would be a continuous role and not one to be exhibited at a single point of time. The planning unit at the district would have to take note of those and incorporate them into their block and district plans, which they would be modulating from time to time. The committee therefore concluded that the district plan was not merely an exercise in producing a participant plan or a long-term action plan. It should be an exercise also in continuous formulation of sound projects and micro unit plans.

Recommendations of the Dantwala Working Group on Block-level Planning

Though constituted *prima facie* for preparing guidelines for block-level planning, the working group did not confine itself to that narrow and specific issue. It was not an isolated exercise on area planning at a particular level. The working group had made it very clear from the beginning that district and block-level planning could be visualised only as part of the same exercise. The remoteness of the planning agencies from the area of implementation and the vastness of geographical coverage had been hampering the matching of sectoral financial allocations with location-specific needs as well as potential and getting in the way of proper distribution of the gains of development. The main purpose of block-level planning was to accelerate the process of decentralization in the hope that a more systematic planning effort with a smaller coverage would not only help to speed up the development process but also improve it qualitatively in the sense of making it more responsive to the needs of the community.

In the opinion of the working group, block-level planning was not to be viewed as an isolated exercise but as a link in a hierarchy of levels from a cluster of villages below the block level to the district, regional and state level. The working group, which had submitted its report before the Asoka Mehta Committee had finalized its recommendations, recognized the block as a unit of planning. The block was distinguished by a certain community of interests. It was sufficiently small in terms of area and population to enable intimate contact and understanding between the planners, those responsible for implementation of the plan, and the people. It could also provide an observation platform in close proximity of the beneficiary group and help to understand more clearly the felt needs of the people and factors inhibiting the uplift of the weaker sections, to aseertain area-specific physical and human resource potential, to identify constraints inhibiting socio-economic and technological growth and to expand the area of people's participation in the preparation and implementation of plans.

According to the working group, the range of planning functions at the block level was determined by the size of its geographical limits and a quantitative and qualitative appraisal of its human and material resources. One of the first tasks of block-level planning should, therefore, be to demarcate the block-level planning canvas from that of the district and state-level functions. The plan for the block had, therefore to be related, on the one hand, to the plans of the districts and the state levels and, on the other, to the plans for the sub-units consisting of cluster of villages within the block. The working group gave only an indicative idea of activities which could be planned and executed at the block level without sacrificing planning efficiency. These could be—

(i) Agriculture and allied activities ;
(ii) Minor irrigation ;
(iii) Soil conservation and water management ;
(iv) Animal husbandry and poultry ;
(v) Fisheries ;
(vi) Forestry ;
(vii) Processing of agricultural produce ;
(viii) Organizing input supply, credit and marketing;
(ix) Cottage and small industries ;
(x) Local infrastructure ;
(xi) Social services :
  (a) Drinking water supply ;
  (b) Health and nutrition ;
  (c) Education ;
  (d) Housing ;
  (e) Sanitation ;
  (f) Local transport ;
  (g) Welfare programmes ;
(xii) Training of local youth and upgrading of skills of the local population.

There would be certain other linked activities which would have to be undertaken as a part of larger resource development programme cutting across block boundaries. Each concerned block could develop a preparatory plan to speed up the absorption (or utilization) of the benefits of those projects. Giving an illustration, the working group mentioned that in the case of a new irrigation project the block-level plan could take up the complementary programmes of land shaping, distributory channels, crop planning, agricultural extension and supply of inputs and credits. Such a complementary planning programme required not only identification of the needs for various inputs but also a proper sequencing of activities. Failure to visualize linkages with programmes of larger area level could cause various bottlenecks. Subject to those considerations, the block-level plan would consist of programmes of intensive development of agriculture and allied activities and provision of the related economic and institutional infrastructure through a large number of programmes oriented to the needs of target groups of landless labourers, small and marginal farmers, rural artisans, women and children.

While the development programmes should aim at the creation of productive assests, they would have to be so designed as to generate maximum possible employment. The block plans and their implementation should, in fact serve as a laboratory for generating through the process of trial and error and experience, a model which could be adopted for achieving the goals of eliminating poverty and unemployment.

The working group had given careful thought to the question of establishing planning machinery. Some measures had been adopted during the Fourth and Fifth Plans to strengthen planning machinery at the state level. In a majority of States, no attempt had, however been made to improve personnel resources at the district level. Even in the case of States which had set up district planning units, no serious attempt had been made to induct technical skills in planning. In the State of Maharashtra, which has a district planning structure consisting of one Deputy Chief Executive Officer (Planning) and some supporting staff, the component of the technical expertise was still deficient. At the block-level, the existing staffing pattern was tuned to implementation tasks rather than planning functions.

Before recommending the kind of organizational structure that would be necessary for undertaking the work of planning, the working group gave some thought to the functions that emerged as pertinent in decentralized planning at block and district levels. Those would include :

(i) Resource inventory and data collection, including analysis of the prevailing levels of development, potential for further development and identification of constraints in development;

(ii) Identification of priorities and selection of a catalytic programme ;

(iii) Formulation of programmes/projects/schemes for development and the establishment of their spatial and temporal linkages within an integrated framework ;

(iv) Devising a plan for fuller utilization of manpower resource ; in other words, a plan for augmenting employment ;

(v) Assessment of availability of financial resources from various sources—district budget, banking system, private sector —and mobilising them ;

(vi) Monitoring and concurrent evaluation or parallel audit of development plans and their modification from time to time in the light of experience.

The working group came to the conclusion that the planning machinery available at the district and block levels was not suitably equipped to perform all those technical functions. It was deficient both in expertise and staff strength. Taking into consideration the realities of the situation, including shortage of technical planning personnel and financial constraints, the working group expressed the view that at the current stage of development it would be better to pool the available planning skills at the district level instead of scattering that scarce resource over several blocks. If the district planning cell could be strengthened suitably, it could take care of block-level planning also. Apart from ensuring the optimum deployment of the personnel resources available for area planning, it would also enable the planning team to take a broader perspective of resources potential and constraints and ensure better co-ordination and integration of the block plans with the district plan. That process would ensure that the block plan was in harmony with the district plan. The working group visualized the planning team located in the district, essentially as a peripatetic team, which would move down to the selected blocks and prepare the block-level plans in association with the BDO, the *Panchayat Samiti*, voluntary agencies and others concerned at the block level. The block plans prepared by the team would be fitted into the district plan.

While the technical task of planning itself should be performed by the technical team constituted at the district level, some part of the planning function could be assigned to the block level. That would include collection of certain types of data and assessing public opinion on various alternative solutions. Generally, the BDO and his staff should be advised by members of the elected *Panchayat Samiti* organizations of the rural poor and other disadvantaged groups.

After giving due consideration to the existing organizational structure at the district level and the skills available, the working group came to the conclusion that areas that would need strengthening in the current machinery for planning would include resource inventory and analysis, project formulation and appraisal, manpower planning and budgeting, monitoring and evaluation and credit planning. The working group recommended that there should be a professionally qualified team at the district level under a Chief Planning Officer who himself should be a person who could co-ordinate and guide the entire planning team under him and also co-ordinate the work of the planning team with that of the departmental officers. He should be the principal aid to the District Collector and next to him on the planning side. He could be from any discipline but his status in the district hierarchy should be such as to enable him to play his role effectively. The Chief Planning Officer should be assisted by the following core staff of technical officers :

(a) Economist/Statistician (project formulation and evaluation);
(b) Cartographer/Geographer;
(c) Agronomist;
(d) Engineer (irrigation/civil);

(e) Industry Officer (small and cottage industries);
(f) Credit Planning Officer.

The role of the *Panchayati Raj* institutions in the planning exercise was also given due consideration by the working group. They had no reservation on that point and stated in unequivocal terms that active involvement of *Panchayati Raj* institutions in the planning process would provide a better climate for people's participation in the implementation of the plan, a condition crucial to its success. Unfortunately, the position was that, except in a few States, there was hardly any worthwhile organization of *Panchayati Raj* institutions. In several States, where such institutions had been formerly established, they had either withered away or were in a moribund state. The committee had also made a pointed reference to the dominance of *Panchayati Raj* institutions by vested interests and to the necessity of providing adequate safeguards to counteract that. There was a general feeling that weaker sections of the rural community did not feel that their interests would be fully protected under the *Panchayati Raj* institutions. Some even felt that they often acted as "gate-keepers" and prevented the flow of benefits to the weaker sections of the rural community. The committee had also hinted at the possibility of a form of local government which would satisfy the democratic aspirations of the elected representatives and give them an executive role, at the same time giving an important place to the administrator.

The working group also discussed the role that could be played by voluntary agencies and professional institutes in the preparation of block-level plans. The rationale for entrusting the task of preparing block plans to voluntary agencies was that they had a dedicated band of workers well-equipped with the felt needs and aspirations of the people in their area of operation and as such they would be in a better position to obtain people's co-operation in the implementation of development programmes. Planning at levels, however, involved a good deal of technical expertise. Though all voluntary agencies did not possess this in the requisite measure, some of the better known voluntary agencies had either acquired such competence or were in a position to hire it as and when required.

The working group was of the view that entrusting the task of formulation and/or execution of some distinct sectoral plans to a voluntary agency would be a step in the right direction. Their plans should, however, be integrated with the district and block plans.

The idea of inviting some professional institutes with necessary expertise to assist the state governments in the preparation of block plans was welcomed by the working group. A list of such institutes could be prepared by the Planning Commission and made available to the state governments. The institutes would, however, be well advised to work in close co-operation with the officers of the state government who were expected to know the ground, the operational problems and the factors inhibiting progress. In the opinion of the working group, formulation of a plan and its implementation could not be dichotomized and those who had to implement a plan should be closely associated with its formulation and vice versa.

Recommendations of the Sivaraman Committee on the Role of Voluntary Agencies

The committee was of the view that voluntary agencies could play a very useful role in planning and implementation of the area development plans. It would be preferable for the voluntary agency which prepared the plan to be involved in the implementation as well. Even in the preparation of a block plan, voluntary agencies with sufficient standing in the practical planning of area development could play an effective role. Some voluntary agencies might not be able to shoulder the responsibility for a block plan. They could be entrusted with the preparation and implementation of a part plan also. Preparation of master plans, being very specialized work, should be entrusted to voluntary agencies (in addition to academic institutions and good management consultancy organizations) having the necessary expertise, proven standing, practical experience and well-equipped organization. Some agencies, such as the Association of Voluntary Agencies for Rural Development (AVARD) had already done considerable work in that field. In the massive programme of rural development which was now envisaged, the committee was in favour of encouraging and supporting other such institutions which could take up preparation of plans for integrated rural development works and provide guidance in their implementation.

The Emerging Picture

Taking into consideration the recommendations of the three high-level committees referred to in the preceding paragraphs, the picture that emerges could be briefly summarized as follows.

In a country such as India with its vast size and natural resources distributed unevenly among the states, regions and districts, it is not possible at the national, sub-national or regional levels to take into consideration the diversity in the levels of development and living, and make realistic plans taking into account potentialities for development and the resources available. The remoteness of the planning agencies from the areas of implementation has all along hampered the matching of sectoral financial allocations with location-specific needs. The kind of planning that has been attempted so far at the national level and at state levels has contributed to regional and social imbalances. Though certain area-based plans and beneficiary-oriented programmes were introduced to correct these imbalances, this can be called only a small beginning. During the Sixth Plan, it is necessary to take such developments to their logical ends. Local-level planning or area planning has to be more effectively introduced in order to achieve this objective.

The democratic process cannot terminate at the state level. Experience has shown that people's own representatives at the grass-roots level are best suited to turn the soil in rural India and induce social change. It is, therefore, imperative that democratic institutions at all levels should play an important role in the planning and implementation of rural development programmes.

The district will have to be the first point of decentralization below the state level. The *Zilla Parishad* will have to play an important role in the planning and implementation of all development programmes of the district in those states which decide to set up such democratic institutions at the district level.

Establishing a district-level body with limited responsibility or having two rival organizations at the district level will be an unwise and retrograde step. No serious attempt has been made so far to induct technical skills in planning at the district level. It is necessary to have a strong planning unit at the district level under a Chief Planning Officer assisted by competent experts.

Preparation of block-level plans will be a step in the right direction. Block-level planning should not, however, be viewed as an isolated exercise but as a link in a hierarchy of levels from a cluster of villages below the block level to the regional, district and state levels. District and block-level planning should be visualized as part of the same exercise. The main purpose of block-level planning is to accelerate the proccss of decentralization in the hope that a more systematic planning effort with smaller coverage will not only help to speed up the development process but also make planning more responsive to the needs of the weaker sections of the community. The district plan will be a resource development as well as an infrastructure plan for supporting the production-*cum*-employment thrust of the block-level planning. The infrastructure required for the various block-level plans and the location of the institutional centres will have to be worked out for optimum support to the block-level plans. The district plan will also take into consideration the institutional supports available, input services and the marketing facilities that have to be developed. The block-level planning will have to take note of the local resources, manpower, institutional strength and public participation for evolving a suitable programme relevant under local conditions, meant to cover the target groups. Programmes suited to local environment and local capacity will have to be identified while preparing the block plan. This exercise can be done more effectively at levels below the block.

It will be apparent from what has been stated above that the planning process cannot stop at the block-level. The *Mandal Panchayats* representing a group of a villages will be able to collect and supply necessary information about various schemes relevant under local conditions and suggest potential areas of development and type of requirements that have to be met. There should be a continuous dialogue between the *Mandal*

Panchayats and the block-level planning organization in the planning process. While it may not be necessary at this stage of development to establish a unit of planning in each block, the administrative machinery will have to be suitably overhauled and strengthened to meet the requirements of block-level planning. Though it has been generally accepted that the Block Development Officer will play the co-ordinating role, there is no functionary at the block who can at present shoulder the responsibility of getting a block level plan prepared even with the assistance of an expert team. Experience has, however, shown that the block administration can be of immense assistance in the preparation of an over-all development plan for the block.

It has to be kept in mind that what is contemplated is micro-level and multi-level planning. While the process of planning is meant to be decentralized, each local-level plan will have to be fitted into the plan at the higher level and ultimately into the state and national plans, keeping in view the objectives as accepted in the national and state plans. In order to do this effectively, there should be some uniformity in the process as well as in format.

Significance of Local-Level Planning in the Sixth Five-Year Plan

Elimination of unemployment and significant under-employment along with an appreciable rise in the standard of living of the poorest sections of the population has been embodied as one of the primary objectives in the draft Sixth Five Year Plan (1978-1983). Emphasis has also been laid on the provision by the state of some of the basic needs of the people, such as clean drinking water, rural roads, rural housing for the landless, elementary education, adult literacy and proper health care. Norms of minimum needs have been specified in respect of all these items. The strategy for rural development in the Plan will be integrated rural development.

The Government has accepted that in the next phase of development it will no longer be appropriate, in the light of past experience, to formulate the principal objective of a particular plan period merely in relation to a specified target of growth for economy. What matters is not the precise rate of increase in the national product that is achieved in 5 or 10 years but whether within a specified time frame a measurable increase can be ensured in the welfare of the millions of poor. It has also been decided that a serious attempt should be made to provide full employment to all the rural population within a time frame of 10 years (to end 9 years from now). To achieve the employment objectives of this Plan, development programmes have to be made area-specific to a much greater extent than in the past. It is, therefore, necessary to have a clear concept of what area planning implies and what lessons have been learned from the experiments conducted so far. Luckily, the operation is not taking place in a vacuum.

The area development should be both area and problem-specific in the first instance, and should also lay adequate emphasis on people, considered both as the object and subject of planning. It has often been noticed that an area development plan leads only to area prosperity and not to people's prosperity. While it is accepted that area prosperity will also bring prosperity to the people, it so happens that it brings prosperity to the wrong people and not to the target groups. Programmes like SFDA have contributed to the prosperity of target groups. What is, therefore, necessary is to interlink the beneficiary approach with area development programmes. This is what is proposed in the programme for developing 3,500 blocks in the course of the current Five-Year Plan.

With regard to recent experiences in the field of micro-level planning, while preparing for the seminar on block-level planning, the Planning Commission had made an attempt to review a number of block and district level plans prepared by voluntary agencies, academic institutions and government organizations with a view to gaining some insights into the various methodologies adopted in micro-level planning. These included:

(i) Block-level studies/plans prepared by voluntary agencies like AVARD;
(ii) Studies made by academic institutions like the Indian Institute of Management, agricultural universities, etc.;
(iii) Block-level studies made by the Pilot Project on Growth Centres;
(iv) Block-level studies/plans prepared for the Pilot Intensive Rural Employment Projects;
(v) Project reports prepared for the Comprehensive Area Development Programme (CADP) in West Bengal;

(vi) Credit plans prepared by lead banks.

Among the studies/plans reviewed, the nature, content and objectives of the exercises varied over a wide spectrum—from a fact-finding survey, through a problem-specific and action-oriented study to an area development plan. So far as area development plans were concerned, it was found that many of them were tailored to the individual requirements of the planning agency concerned and did not deal with the totality of the development requirement of the area. The plans were not related to resource availability and there was no prior commitment that the plans prepared would be taken up for implementation. However, in many cases, it appeared that a lot of valuable data has been collected which could be a starting point at the time of preparation of a comprehensive area development plan.

A specific instance may be given of the plan made by AVARD for Adhaura Block in the Rohtas District of Bihar. The CADP of West Bengal had divided planning into two phases, confining the first phase of work to planning at the micro level and the second to more detailed planning with greater involvement of the local people. Even in this case, it will appear that detailed plans prepared with the help of certain expert bodies and consultants did not always correlated the plan to available resources and existing realities.

A plan prepared for the "Whole Village Development Programme" in 105 villages in Mirzapur District by AGRINDUS Institute, Banvasi Seva Ashram in the District of Mirzapur, has been an extremely useful and painstaking exercise. A survey was made of all the families in 5 villages and of 10 per cent of the families in 100 villages. Another project carried out by the same organization attempts to take advantage of the maximum potential of village industries in the area.

The study conducted by the Planning Commission came to the conclusion that in the current situation, the data limitations were bound to be a constraint for planning for a long time to come. Since, however, planning cannot wait and the very "best" analysis may not be worth all the trouble if it comes too late for the decision, it was felt that some satisfactory solutions will have to be found through "quick inferential procedures", "random sampling, techniques" or "remote sensing techniques" wherever possible. Such approaches, even if they imply sub-optimum planning; would be preferable to pure guesswork. This is an aspect of the theory of limited rationality which aims at "satisfactory" solutions rather than "optimum" solutions. Many developing countries have recognized the need for such "reduced approaches" in planning. So, in block-level planning too, appropriate methods have to be devised of collating available information through interpolation and extrapolation techniques.

In consonance with this approach, attempts have been made recently to find out the kind of data that are available or that could be made available at present. An attempt had been made by Karnataka State to collect essential plan statistics for each of the blocks in a district to be supplemented by corresponding village-wise plan statistics. The state government had considerably strengthened the planning machinery at the state level by establishing a District and Regional Planning Unit attached to the Economic Adviser's Division. This Unit collected block-wise plan statistics in a meaningful way for a number of years. A state like Maharashtra which had made a lot of progress in local-level planning had also been able to collect a lot of useful statistics which could serve as a starting-point for block-level planning. Admittedly, there are many gaps but these can be closed as the process of planning progresses. Block plan statistics are based on various statistical reports and studies. These data refer to different periods of time because they are based on some field surveys—quinquennial surveys and decennial surveys. In the resource inventory, it will be necessary to fill in the figures on the basis of the latest information available and indicate the year of information suitably in the format.

In a programme of integrated rural development meant to cover 20 selected districts, the Indian Council of Agricultural Research had, through the agency of its scientists or agricultural universities, tried to make a detailed resource inventory of the districts. Though detailed data have been made available in respect of a few districts, it requires expertise of a different type to interpret these data and suit them to local requirements. Detailed data alone will not help. What is needed is adequate and purposeful data on which an area plan can be framed. Resource inventory will have to be based on a family-based programme as has been done in a small project at Somangalam in Tamil Nadu. The methodology adopted for planning and monitoring this scheme could be advantageously utilized in area development programmes at the block level.

A committee constituted recently by the Planning Commission to evolve the basic frame for the guidelines for block-level planning visualizes the following steps in the preparation and implementation of block-level plans.

Assessment of Resources

Resources for block plans will include state plan outlays, outlays for the Minimum Needs Programmes (MNP) of the states, sectoral outlays that may flow from central funds for various special programmes, special central assistance for block planning and institutional finance. So far as state plan outlays are concerned, the share of resources that may accrue to the block from the divisible portion of the state plan has to be assessed. The first step in this direction will be for the state to communicate the resources that would be made available to the district on the basis of some rational approach, giving due weightage to the relative levels of development among the districts. The district, in turn, will have to make allocations to the blocks, taking into consideration the needs and levels of development of different blocks. Experience in the states of Maharashtra, Gujarat and Karnataka has indicated that 30 to 40 per cent of the annual plan outlay could be available for distribution among the districts. Further disaggregation of the outlay from the district to the block will have to follow a broad assessment of the requirements for those coming under the district sector schemes and those that could be allotted for block sector schemes. So far as Minimum Needs Programmes are concerned, an assessment will have to be made of the shortfall in each MNP item in the area for which the plan is being prepared. This is likely to be a difficult exercise; but some kind of rationale will have to be developed as an interim measure. So far as central sector scheme outlays are concerned, the quantum available for each district is easily ascertainable. Divisible resources under these schemes will have to be allocated blockwise. Over and above this, special assistance is being given blockwise for the intensive programme. All these things will have to be taken into consideration while preparing the block-level plan. It may again be emphasized in this connexion that block-level planning is not expected to be an isolated exercise. It has to be correlated to the district plan which, in turn, is interlinked with the state plan. Requirements for institutional finance will have to be separately assessed, taking into consideration the requirements of various beneficiary-oriented programmes as well as the general requirements of the area. A separate exercise will have to be conducted at the district level to ascertain the institutional support already available and the input services and marketing facilities that have to be developed.

Ascertaining the Scope and Functions of the Block-Level Plan

An indication has already been given of the activities which should be considered relevant to block planning.

Preparation of a Resource Inventory

A resource inventory should not be a mere shopping list. It does not help to have too much data, as mentioned. What is needed is adequate and purposeful data on which an area plan can be framed. An attempt is being made by the committee to identify the critical minimum data and information needed for block-level planning, taking into account some basic work done by various state governments, agencies and experts. A final pronouncement is yet to be made on the adequacy or otherwise of these data.

Formulation of Projects

Guidelines will have to be given from time to time for preparation of projects suitable for meeting the objectives. Guidelines can, however, be only illustrative. Projects relevant to the area will have to be selected according to the local needs and aspirations of the people and the resources available.

Household Planning for Families

Detailed household planning will have to be made in respect of the target families within a definite time frame. A format will have to be developed for this according to the needs of the programme. The idea is to keep an eye on the economic condition of the family and to make sure that the targeted families are raised above the poverty line according to plan.

Spatial Planning for Socio-economic Infrastructures
This will have to be dovetailed into programmes relating to economic development. Since infrastructure development plans will have to be formulated mostly at the district level, this will have to be co-rrelated to the district plan.

Arrangements for Staff and Machinery for Block-level Planning
The Dantwala working group recommended the setting up of a planning cell at the district level under a Chief Planning Officer with a minimum core staff of six technical officers, as previously mentioned. In addition, specialists may have to be engaged according to the needs of the area or the programme. Though a separate planning cell is not needed at the block level, the block administrative machinery will have to be suitably strengthened. The training and visit system envisaged for agricultural extension work may take care of agriculture development programmes. The role that is to be played by each functionary will have to be properly defined. It will also require additional staff, both on a temporary basis to do surveys and on a permanent basis for implementation and monitoring. If the progress assistants posted at the block headquarters are given proper training, they can play a useful role in the collection and compilation of data as well as in the monitoring programme.

Training
A training programme will have to be organized after the parameters of block-level planning are prescribed. The National Institute of Rural Development can play an effective role both in developing the methodology of block-level planning and in the training of trainers.

Progress Monitoring
Arrangements will also have to be made for monitoring and review of progress. Monitoring and evaluation cells will have to be established both at the state headquarters and in the districts. Apart from evaluation that is entrusted to specialized institutions and consultants from time to time, a built-in

system of evaluation and an efficient monitoring system are absolutely essential for the success of the programme.

CONCLUSION

In the integrated rural development programme now under implementation, the block has been accepted as the unit of planning and implementation. Block-level planning will, no doubt, accelerate the process of decentralization, speed up development and make programmes responsive to the needs of the target groups belonging to the weaker sections of the society. It should, however, be remembered that block-level planning cannot be thought of as an isolated exercise but should be fitted into a hierarchy of levels both above and below. Micro-level planning cannot end at the block. At the same time, the other is in agreement with the finding of the Asoka Mehta Committee that a village is too small a unit either for the organization of local administration or to serve as a unit for planning. Micro-level planning will have to be done on the basis of a cluster of villages or the hinterland of a growth centre. Block-level plans will have to be prepared taking into consideration materials collected from these smaller units. Planning has to be both micro-level and multi-level.

There has been some adverse criticism of the selection of the block as a unit for planning. Since programmes can be planned and implemented much more effectively on the basis of a command area or watershed, some experts have questioned that propriety of making the block itself a unit for planning.

While they accept the district as a unit for local-level planning, they feel that micro-level planning should not be done at the block-level. Agreeing fully with the point of view expressed by the Dantwala working group that district and block-level planning are to be visualized as a part of the same exercise, the author is of the view that the block can, in the present context, be taken as a convenient administrative unit for planning and implementation of all rural development programmes. This need not necessarily clash with the concept of watershed management and command area development. Where a command area or a watershed cuts across a number of blocks, proper co-ordination can be ensured at the district level. The point to be emphasized is that block-level planning is not an exercise in isolation.

If block-level planning is to be successful, all organizations concerned in the process will have to get closer together. These will include the administrative machinery at the block level, the *Panchayati Raj* institutions at block and village or *Mandal* level, the co-operatives and other financial institutions, as well as voluntary agencies and academic bodies involved in the process of area development. An atmosphere is now being created in which government machinery is willing to extend full co-operation to non-official institutions and voluntary agencies. The author does not agree with the statement that entrusting the task of preparing an area plan to an outside agency will imply the expression of a lack of confidence in the ability of the government machinery. The Government has, off and on, taken the assistance of experts and consultants in preparing plans and project reports. There is no reason why such expertise cannot be introduced in the rural development programme in the preparation of block-level plans, sectoral plans or even district plans.

One valid criticism against the present programme based upon micro-level and multi-level planning could be that it is an attempt to go too fast before gathering sufficient experience and expertise. Two thousand out of a total of 5,004 blocks in the country have been selected for a programme of intensive development based on block-level planning. In addition, 300 additional blocks are scheduled for intensive block-level planning and development every year during the plan period starting from the current year. Voluntary agencies and academic institutions with expertise in the field may not be able to handle more than 100 blocks during the first year. This means that the normal administrative machinery, with the help of *Panchayati Raj* institutions wherever they exist, will have to bear the main burden. Doubts are being expressed whether they will be able to shoulder this new responsibility.

On the other hand, there is equally strong criticism that large areas of the country are not getting the benefit of the new integrated rural development programme. Taking into consideration both aspects of the case, the Government had to take a balanced view and decided on a phased out programme. By

1983, 2,000 blocks will have had the full impact of the programme for a period of five years and other blocks will be at different stages of development. It may not be possible to have detailed block-level plans for most of these blocks at the initial stage. The programmes will have to be perfected as we proceed. In this process, it is possible that resources might have been stretched to the utmost, but the situation surely demands it.

TWELVE

INDONESIA

National Institute of Public Administration, Jakarta

Introduction

SINCE the Government started the First Five-Year Development Plan in 1969, all development programmes have always put strong emphasis on the development of the villages or rural areas. This is based on the fact that 82 per cent of the Indonesian population live in the rural areas. They are classified as the low-income groups of society. By developing the villages, the quality of life of the majority of the population of Indonesia will be improved. Therefore, a great deal of effort—sectoral, regional as well as Presidential Grants in Aid (INPRES) projects—is directed toward the development of the rural areas.

There are 60,445 villages in Indonesia, the development stages of which can be classified into three main categories :

(a) The *Swadaya* village, described as a traditional village or society in which economic conditions are only sufficient to fulfill the people's primary needs;

(b) The *Swakarya* village, which has been exposed to external influences, including the introduction of appropriate technology;

(c) The *Swasembada* village, described as a more advanced and modern village that will be further developed later as a *Pancasila* village.

The *Pancasila* village is the ultimate goal of the rural development programmes as the realization of ideas stated in the 1945 Constitution and the "Broad Guidelines of State Policy". Such a village can be described as modern and advanced within which all the basic needs of the people are fulfilled on a self-sufficient basis such as : food, clothes, health, education, employment opportunity, security, maintenance of law and order, and social welfare benefits will have been equally distributed among the members of the community.

The Government has decided that within 25 years from 1974, all villages in Indonesia should have developed into *Swasembada* villages. It was expected that the rate of the transformation would be 4 per cent yearly. Experience has indicated, however, that so far the rate of development has been only 1.5 per cent yearly. Therefore, efforts are now being launched to find a new system or technique of development to accelerate the rural development target rate.

Problems in Rural Development

One of the important objectives of the Five-Year Development Plan is to improve the standard of living of the rural population. Various kinds of sectoral policies have been adopted to solve these problems by implementing several kinds of development programmes followed by the advocation of several concepts of rural development planning.

The problems faced in the development of rural areas can be classified into three major problem areas :

(a) A problem of ecosystem, which is characterized by uneven distribution of population and settlement. There are three patterns of population distribution and settlement in Indonesia :

(i) Java and Madura are inhabited by about 65 per cent of the total population of Indonesia, while covering only 7 per cent of the total land area (about 1,000 inhabitants per km^2). In those islands, natural resources have already been exploited to their maximum and have almost reached critical points of utilization. Job seekers in the villages are increasing while employment opportunities cannot accommodate job demands, which results in an increase of unemployment in the villages;

(ii) The islands outside Java and Madura which are sparsely populated (about 20 inhabitants per km^2).

Mostly, their natural resources have not been exploited yet, primarily owing to the lack of man-power and infrastructure;

(iii) Villages where there is a proper balance between the distribution of population and their natural resources, but which have not been managed and developed yet.

(b) The second major problem is the fact that in most villages, the social conditions are still below standard. These problems include health, which is still far from satisfactory, low level of education, inadequate housing facilities, and low productivity, that results in the low level of income of the community and its welfare;

(c) The third major problem is the unsound management system. This problem is characterized by the ineffective functioning of the institutes in the villages, scarcity of dynamic, creative and fair leadership, inefficient co-ordination of development efforts, low quality and quantity of village administrative machinery, inability in planning, weakness in implementation, and lack of efficient control of the implementation activities of the development programmes.

Local Development Working Unit

Recognizing the lack of capability of the existing machinery in the villages to reach the set target in rural development, the Government is searching for a new and more effective system to speed up the development of rural areas. Besides, the Government realizes the different social and economic stages of the rural areas as well as their interdependency and their mutual symbiotic nature. These conditions contribute to the effort to find a new system of developing villages in Indonesia.

The main purpose of development of the new system in promoting faster growth in the rural areas is to speed up the attainment of *Swasembada* village status as a stepping-stone to the achievement of *Pancasila* village status.

One of the basic principles underlying the new system is that the development of rural areas is a subsystem of regional development and at the same time it is a sub-system of national development. To make this system approach more meaningful, since the beginning of the implementation of the Second Five-Year Plan in 1974 a better concept and system of regionalization has been developed. All the development centres throughout the country have been categorized as follows :

(a) Main development areas with main development centres (4);
(b) Large development areas with development centres (10);
(c) Medium development areas with development centres (85).

The principle of regionalization of development planning serves only as a "mapping" guide. In reality the implementation of the development plan is always performed on the basis of administrative territorial considerations i.e. national, provincial, *Kabupaten* (municipal) and *Kecamatan* apparatus with their villages. Based on this system of regionalization, the new method considers the ecosystem and the potentialities of each development area which can be developed and utilized for the identification of smaller development areas subordinate to the *Kabupaten* development area. *Kecamatan*, as a rural development planning area, which covers a rural ecosystem and has homogenous potentialities, is considered a rural development centre.

The innovated system of rural development is called the Local Development Working Unit (better known by its Indonesian acronym, UDKP). It is an integrated part of the national development system using *Kecamatan* as the basis of operation.

The main functions of UDKP are to establish priorities of rural development, to integrate the sectoral, local, INPRES, and rural community development programmes, and to coordinate and monitor the implementation of the programmes. The *Camat*, as the lowest deconcentrated government official in the region, is responsible for implementation of the rural development programmes in the UDKP so as to maintain the concept of unity of command.

In its operation, UDKP has the responsibility of stimulating the rural community to increase their effort in the primary, secondary and tertiary economic sectors. In addition, UDKP should also provide services to the rural community, such as : consultation, communication, administration and socio-economic services which are appropriate to their dynamic stage. For that purpose, UDKP operations will cover the following activities ; survey undertaken in formulating plans, offering training opportunities to develop cadres by increasing their skills, and improving job methods by introducing appropriate technology.

UDKP Planning

Basis of UDKP Planning

The development activities in an area can only be carried out smoothly and harmoniously when there has been an integrated and comprehensive plan of the region. Based on this plan, all the government agencies in the region and the respective community will programme various development activities according to the functions and responsibilities.

To make a sound regional plan, an accurate and in-depth knowledge of the respective region is essential. This knowledge is in the framework of regional analysis, the role and function of all elements contributing to the establishment of the region, including their role and functions as an integral part of a large region that is capable of developing the community into a more advanced stage.

In formulating UDKP plans, the following aspects should be considered :

(i) The typology and potentials of the respective region ;

(ii) The location and function of a region in its relation to a larger region of development, i.e. the *Kabupaten* as a development region ;

(iii) Other development activities that have been determined at the *Kabupaten*, province, and national levels ;

(iv) The role of *Kecamatan* that will be developed in relation to a *Kabupaten* region.

Knowledge of the Region

To prepare a sound plan, knowledge of the region should be mastered. It should analyse systematically the available resources in the region and cover the following aspects :

(i) Potentials of natural resources and the living environment ;

(ii) Manpower potentiality ;

(iii) The outcomes of the work and technology.

For the purpose of acquiring these data, a survey should be conducted and the data used as a basis of drawing up the UDKP plan.

Preparing a UDKP master plan requires much time and it should be done in several stages. The first stage is to conduct a study on the existing conditions of the physical/natural resources situation and its environment, socio-economic condition, and what could be further cultivated and developed.

The second stage is to identify and analyse problems, identify the proper solutions and draft required plans in line with the pattern of development of the region concerned.

The third stage is to prepare programmes based on established priorities. From the master plan, projects are drafted based on predetermined programmes and formulation of the yearly budget for the projects.

Regional Chart

A regional master plan of development should be completed with information and charts, including : situational chart, analytical chart and plans chart.

Adaptation of Plans

The subsequent step is the implementation of the plan. The UDKP plans become the main subject of discussion at workshops at the *Kabupaten*/municipal level in order to get the response of local government agencies and other government agencies from the viewpoint of their functions and responsibilities. The stages of the planning process, programming and drafting the projects through UDKP can be described as follows :

(i) An inventory is prepared of all projects that are financed by central and local government—province and *Kabupaten*, and the participation of the community ;

(ii) On the basis of the UDKP survey, priorities of development at the *Kecamatan* level are established and then harmonized with the projects decided upon by the *Kabupaten*/ municipal region of development ;

(iii) The projects decided upon should be spelled out in further detail, including their sites, objectives, targets, activities, financing, and time schedule for implementation ;

(iv) The formulation of projects should bederived from workshops at the *Kabupaten*/municipal level and then submitted to the Governor, copies being sent to the local planning board and government agencies in the province. Meanwhile, the decision on projects derived from the *Kabupaten*/municipal workshops which are already clear and concrete in their source of financing and approved by the government agencies can be directly implemented. The discussion at UDKP/*Kecamatan* level is meant to operationalize the implementation of projects which have already been approved at the *Kabupaten*/municipal workshop and to increase community participation in the implementation of those projects ;

(v) The UDKP province workshop, based on adopted national and regional development policies, is to analyse the results of the *Kabupaten*/municipal workshop and to evaluate relative urgency for the purpose of synchronizing with the adopted policies. The Governor of the province will then submit the results of the UDKP workshop to the Minister of Home Affairs, through the Directorate General of Rural Development, and to the Chairman of the National Development Planning Board ;

(vi) Based on the results of the UDKP workshop in the province, the government agencies in the region, on the basis of their respective functions and responsibilities, will include those approved projects in their own programmes. After they have been co-ordinated with the local planning board they are then submitted to departments and non-departmental agencies, to the Department of Home Affairs, and to the National Development Planning Board, for further processing especially through national consultation of the local planning board and the National Planning Board and departments in the Central Government, and during conferences for co-ordination of rural development.

THIRTEEN

MALAYSIA

L.S. Frederics

Past Experience : A Summary*

IN RECENT years the Government of Malaysia has been very much concerned with rural poverty and regional disparities. The strategies for dealing with such problems have been conducted predominantly on a sectoral basis. For instance, a considerable effort has been made to increase the productivity of rubber small holdings (233,800 households) or rice farming (114,300 households), which together accounted for about 42 per cent of all poor households of Peninsular Malaysia. A similar effort has been undertaken for the development of new land settlements under the direction of FELDA.

Simultaneously, an effort has been made to decentralize the planning process at the state/regional level. Several regional plans have been formulated, for instance for the Jenka Triangle, Pahang Tenggara, West Johore, Johore Tenggara, North Kelantan, and North-west Selangor. Regional planning has been directed at formulating a consistent over-all strategy for regional development, in the context of national policies, thus rationalizing resource use and exploiting each region's potential.

As regards the implementation and combination of development programmes, Malaysia is well known for its "operations room" system which exists at the federal, ministerial, state and district levels. Essentially the "operations room" consists of the basic data (maps, projections, charts) which facilitate monitoring of the progress of programmes and identification of

* Summary prepared by the ESCAP secretariat.

bottlenecks. In this context and with the current desire to integrate agency proposals in area development plans, a more focal role will be assigned to the district officer to facilitate resource mobilization for more efficient area and local development. It is to be noted, however, that the programmes to be implemented at the district level do not usually form a coherent plan but are rather a collection of projects originating from various departments or authorities.

Current Approaches to Integrated Rural Development Planning

A detailed description and analysis is given below of current approaches to integrated rural development planning of the following three "local-level" plans :

(a) North-west Selangor Integrated Agricultural Development Plan (NWS);
(b) National Small-scale Irrigation Projects (NSI);
(c) Krian/Sungei Manik Integrated Agricultural Development Projects (KSM).

These three agricultural plans are examined from the following points of view : objectives; scope; data collection and analysis; programme formulation, implementation and evaluation; linkages with national strategies and popular participation in planning and implementation.

Objectives

The three projects are all directed towards an intensification or improvement of existing irrigation infrastructure to improve water control, flood control and water table control in the rain-fed area, and provide more reliable water supply and drainage in the irrigated area. Thus construction, improvement or rehabilitation of an irrigation system is identified as a major prerequisite to on-farm development to increase farm productivities, aggregate output and farmer incomes.

Each of the schemes spells out the broad objectives to be attained, invariably in terms of raising productivity and incomes for a defined group of farmers (see table below) cultivating rice, either alone or together with other crops. It must be assumed that goals are primarily agricultural development rather than rural development-oriented. In a way, this follows the traditions of planning which specify the formulation of quantitative goals in terms of the target group, area and specific indexes to be realized. An examination of the specific goals in the following table attests to this quantification of goals in terms of new land irrigated (NSI) or better irrigation of old land leading to higher production efficiency, aggregate output increases and better incomes.

It should also be observed that given the variability of the physical or cultural environments obtaining in the project areas, goals are differentiated in terms of main and off-season yields, among non-Chinese and Chinese farmers (in NWS, the Chinese rice farmers of Sekinchan are more effiicent resource users than the Malays and Javanese; the incremental yields that can be realized are therefore higher among the latter two groups than the former). Most important (although lacking in KSM), as an integrated agriculture development plan must consist of an irrigation and agricultural services component, a differentiation in yields is anticipated. Thus in NWS, main season yields can be expected to increase up to 3.7 or 4.4 tons/ha depending on the full implementation or otherwise of the complementary agricultural services component. Similarly, in NSI, a "bonus" incremental output of 0.5 tons/ha can be realized with the full implementation of the integrated package to farmers. The specification of ranges in quantitative goals also indicates the degree of confidence of the planners in anticipating yield increases from technical inputs (irrigation) in a project as compared to its institutional (agricultural services) inputs.

Goal specification is then clear and quantitatively expressed. The specification of ranges in the desired goals indicates either a good planning technique or relatively lower confidence of planners in certain components of the plan package, or both.

Scope

By scope of the plan is meant the instrumentalities by which a plan or project attains the stated goals. Two basic instrumentalities are indicated in the three project plans examined, the technical component and the agricultural services component.

In all the three schemes, improvement, construction or

Goal Specification

Projects	*Major goals*	*Specific goals*
NWS	Raising productivity/incomes of 32,000 rice and tree crop farmers	(a) Improved irrigation for 20,000 ha (rice) and 77,000 ha (tree crops) (b) Increase in: main season yields from 3.0 - 3.7/4.4 tons per ha off-season yields from 3.5 - 4.0/4.7 tons per ha (c) Incremental output from 25,000 - 46,000 tons/year (d) Income increases from: \$US 583 - \$US 735/year/farmer (Malay) \$US 1,100 - \$US 1,150/year/farmer (Chinese)
NSI	Raising productivity/ incomes of 60,000 rice farmers operating outside large-scale irrigation areas	(a) New irrigation and cultivation of 13,000 ha; double cropping of 20,000 ha (b) Cropping intensity from 109% (40,000 ha) - 158% (53,000 ha) (c) i. Increase in main season yields: from 1.9 - 2.8 tons/ha; off season yields from 2.3 - 3.2 tons/ha; (old land) ii. New land yields: 3.8 tons/ha (main season); 4.3 tons/ha (off-season) iii. Incremental output of 0.5 tons/ha if agricultural package implemented (d) Average income raised by \$US 119/farmer

KSM	Improving socio-economic position of 23,600 rice farmers	(a) Improved irrigation for 30,600 ha of rice land; improved agricultural support services for 45,800 ha of other land (b) Increase in yields from 2.9 - 3.8 tons/ha (Krian); 2.5 - 3.5 tons/ha (Sg. Manik) (c) Incremental output of 76,000 tons of rice (d) Cropping intensity from 170% - 195% (Krian); 152% - 190% (Sg. Manik)

rehabilitation of the existing irrigation systems is a prerequisite to improving the agronomic potential of an area.

In the NWS scheme, the early settlers constructed an open drainage system to permit coconut cultivation but the system suffered from an inability to clear localized flooding, sea-water intrusion and river inundation because of jungle clearing. In 1910 and 1932, government drainage and flood control works were initiated leading to the development of a 90 km network of coastal bunds, tidal gates and main drains. Between 1937 and the 1950s, the Drainage and Irrigation Department (DID) constructed an allied drainage system for tree crop agricultural areas, while estates invested in their own drainage systems. However, coastal bunds, river bunds and allied drainage systems are inadequate or require repair and improvement as the area is still subject to inundation and waterlogging.

The irrigation system for rice cultivation in Tanjong Karang was developed from virgin jungle by DID in 1936, by tapping the river and a nearby swamp through the construction of diversion barrages, feeder canals and gates. The existing irrigation system which suffers more from a lack of command and a poor distributory system, has reached its technical optimum in permitting double cropping on approximately 70 per cent of the scheme area. Several deficiencies are apparent, including inadequate capacity of some parts of the main canal, leaky headworks, inadequate command area leading to slow inundation of fields and late transplanting. The total impact has been a situation where large areas of rice are continually out of phase, and drainage for drying and harvesting earlier planted fields is impossible if the water needs of the less mature crops are to be met.

A similar picture reflects the situation in KSM which consists of two areas, Krian and Sungei Manik. The *Krian Irrigation Scheme* was begun in 1906 to provide single cropping of rice from the Bukit Merah Reservoir, but its inadequacy led to the development of the Krian Extension Scheme in 1960 involving three phases. The Extension Scheme resulted in the enlargement of the reservoir and main canal capacities to facilitate main and off-season double-cropping for 30,000 acres, improvement of river flood protection bunds, raising levees and drainage works and a tidal barrage and pumping station to facilitate double-cropping on the remaining 32,000 acres. The *Sungei Manik Irrigation Scheme* was begun in 1933 to reclaim swampy land for rice cultivation and drawing irrigation water by constructing headworks on a major nearby river. Four stages were involved, the final stage being completed in 1952, allowing for single cropping of the entire area. After 1952, improvements to the drainage works and reconstruction of the entire drainage and irrigation infrastructure began resulting in an increasing area under double-cropping.

However, in both areas, inadequacy of water supplies, particularly during the off-season, a slow and uneven distribution system of irrigated water, waterlogging and inundation restrict further increases in the productivity of the region.

NSI, however, is a composite of 195 small-scale irrigation schemes having a command area between 50 and 2,000 ha of net irrigable rice land. The schemes, spread over Peninsular Malaysia, Sabah and Sarawak, are supplied with crude irrigation systems (controlled drainage, inundation systems, and crude gravity systems using brushwood dams) or are dependent solely on rainfall. This results in some areas being uncultivable, insufficient water control required by HYVs (hence, traditional varieties only are used), drainage problems, seawater flooding or lack of road access.

The agricultural services component forms the sufficient condition to exploit the agronomic potential of an improved irrigation area fully. In the NWS project, for instance, the two major components are formulated as follows:

(a) Rehabilitation of the existing headworks and feeder and main canals and structures serving the 20,000 ha of rice fields in the Tanjong Karang irrigation area; construction of required access roads in the rice field area; construction of tertiary irrigation and drainage networks which would serve each individual lot; and provision of required maintenance equipment;

(b) Rehabilitation and improvement of existing main drainage and flood control systems serving 77,000 ha of tree crops, including upgrading and reconstruction to the coastal and river bunds; provision of tidal gates, bridges and gated culverts; construction of an intensive drainage system serving individual small holder lots and farm access roads; and upgrading of the main drainage system in estate areas;

(c) Construction of offices, quarters, storehouses, drying floors, and provision of vehicles, agricultural equipment, and training to support an intensified programme of agricultural extension, agricultural credit, input supply, and marketing services for the region's small holders.

For the NWS project, a whole range of agencies, mostly under the Ministry of Agriculture, provide agricultural support services. These include the Department of Agriculture (DOA), Drainage and Irrigation Department (DID). Farmers' Organization Authority (FOA), Agriculture Bank (BPM), Malaysian Agricultural Research and Development Institute (MARDI), Federal Agricultural Marketing Authority (FAMA), Rubber Industry Smallholders' Development Authority (RISDA) and the National *Padi* and Rice Authority (LPN). The last two authorities are attached to the Ministry of Primary Industry and the Ministry of Trade respectively.

DOA is the key supporting agency as it provides the extension links to rice and rubber small holders and operates the Coconut Smallholders' Development Scheme. Its network of extension agents consist of 20 Agricultural Technicians supervised by 4 Agriculture Assistants and 2 Agricultural Officers. The ratio of over 1,600 farmers per extension agent makes for minimal extension work, particularly as the extension agent is loaded with regulatory duties, statistical and subsidy programme administration.

DID is responsible for O and M (Organization and Management) for the drainage and irrigation system and also advises the District Agriculture Committee (representing the District Office, DOA, DID and farmer groups) on water schedules related to rice cultivation operations.

FOA operates three Farmers' Development Centres which, with 66 co-operatives, provide various farming inputs, marketing and processing facilities for *padi* and non-*padi* crops. Both agencies are not fully mature in terms of management of agribusiness activities or economic viability and hence the role of private enterprises is significant in the area.

LPN operates a *padi* and rice monopoly to buy and process all *padi* produced in the area, aided by licensed buyers and processors. While BPM provides short-term production credit through FOA and Farmers' Development Centres, MARDI operates a small experimental station undertaking research on *padi* and other crops.

The same patterns hold in the NSI and KSM projects in terms of the agricultural support services currently available. For the agricultural services component, a quantitative and qualitative improvement of personnel and outputs is necessary to secure income and productivity gains.

Data Collection and Analysis

Plan formulation, given the critical role of the irrigation infrastructure in improving *padi* output, normally originates on the basis of technical feasibility studies by DID. In the NWS project, the Selangor State DID office initiated efforts to improve the existing drainage and irrigation infrastructure as a basis for intensified cropping in rice-growing areas and intercropping in tree crop areas. Since it was consciously felt that only an integrated package of technical and agricultural improvements would secure an optimum increase in incomes and productivity, once the technical and economic appraisals of the DID component were available, co-ordination with other agencies was programmed. This was facilitated by a Ministry of Agriculture Committee on which were represented the various government agencies involved (state and federal), the state government and district officers. For the NSI project, since it involved a package of 195 schemes, the basic technical work was undertaken by DID which then consulted the Economics and Statistics, Crop Production and Soils Branches of MOA as to the agronomic and economic considerations attendant on improving irrigation systems in several areas. The same pattern of plan formulation as in NWS was followed by KSM.

The data base of each project plan consists of basically three sets of information; largely qualitative data on the strengths and, particularly, the weaknesses of the implementing agencies; socio-economic and other technical data (rainfall, soil conditions, quality of irrigation infrastructure, etc.), pertaining to the project area; and data on which the economic feasibility of the whole project is determined.

DID, as the major implementing and executing agency, has been consistently evaluated as an organization of proven competence in the planning, design and implementation of both large and small projects in irrigation, drainage and flood control. In the case of the NWS project (as in the other two projects), improvement in the staffing position was naturally required. Particularly in the case of NWS, a perceived weakness was the lack of an effective system to bring DID field staff into contact with end-users. A reactivated District Drainage Board and the creation of irrigation committees were recommended to overcome this organizational weakness.

DOA undertakes all extension activities under a State Director of Agriculture who is administratively responsible to the State Secretary, as agriculture is a state matter. The federal DOA provides staff, financing and technical information to the state DOA. While the DOA thrust in agricultural extension has generally been rated as good in the case of NWS, the ratio of field extension agents to farmers is poor, and staff are generally young and inexperienced and have little impact on the farming community. Staff training and additions to field staff are thus necessary if the project is to succeed. A reorganization of the extension services in the KSM project is also deemed a necessity to provide more farmers with this crucial input than is currently possible.

FOA, which supervises Farmers' Development Centres in the project areas, operates through the Area Farmers' Organization (AFO) and its affiliates, the co-operative societies and small agricultural units in supplying inputs and services to members. In NWS, only 7,000 or 22 per cent of all farmers are members of AFO; 50 per cent in KSM; and a small percentage in NSI which covers very poor *padi* farmers who normally are not as well serviced by FOA. Apart from its limited coverage, poor management has been identified as a key weakness leaving its impact, despite its critical role in all areas, with much to be desired. In addition the NSI project document notes the lack of operation and programme planning and evaluation procedures in AFOs. Thus it states that the "..... system requires close review at national level, with a view to overcoming present deficiencies, integrating services with agricultural extension activities and providing effective input supply, marketing and other basic services to all sectors of the agricultural community". In particular, the NWS project plan calls for a judicious selection of agri-business activities; promotion of FOA activities among all farmers and not just members; vigorous training for field staff and an improved organization and management system.

BPM operations are such that little institutional credit is available to farmers because of repayment problems and the difficulties of locating and persuading farmers to repay their loans. BPM operates through designated local credit centers (which are usually AFOs). Given the psychology of farmers who often do not feel an overwhelming moral obligation to repay their loans, a restructured rice production credit system is necessary involving BPM, DOA and FOA working through small farmers' groups. This system has been tested in the North Kelantan Rural Development Project Area and involves AFO as the principal borrower.

In the case of MARDI, two main problems were identified; the need to produce varieties of rice which were location-specific and the need to bridge the gap between research, extension and the farmers. For both problems, the urgent requirement was for the creation of small groups of MARDI research staff working with the state DOA extension service. At the state level, sufficient co-ordination takes place through the Agricultural Research Development and extension committee chaired by the state Director of Agriculture.

The second source of data pertains to the socio-economic and technical information on which to base economic decisions and formulate engineering and management aspects of the scheme. The KSM project paper (drafted by MOA) provides extensive data on land-use and the socio-economic situation pertaining to small-holder agriculture in Krian and Sungei Manik. This information includes land-use patterns (including tree-crop agriculture), structure of ownership of *padi* land (size, tenure, number of parcels), cropping intensity, productivity, marketing, incomes, socio-economic characteristics of farm-household (demographic characteristics, facilities available, rural institutions), cultivation practices and production costs. Since the KSM project is meant to benefit tree-crop farmers also, similar data for rubber and coconut farmers are also provided. Data is drawn from various sources: land-use surveys, 1966 and 1974; a socio-economic survey undertaken in 1976 covering 96 rice farmers, randomly selected from the area; a fruit cultivation survey, and a survey of rubber and coconut small holders, consisting of 48 farmers only. The objective of these surveys is to provide a profile (however inadequate statistically) of the area as well as to indicate problems faced by the farmers. Thus, for example, the KSM annexes on Krian and Sungei Manik point to the poor coverage of DOA in providing extension advice, lack of capital to purchase inputs, insufficient recommendations on input use in different sub-areas and lack of uniformity in planting schedules.

The NWS project provides less socio-economic data although its data base was a survey undertaken in 1976 by the Ministry of Agriculture Economics and Statistics Branch. However, both NWS and NSI provide a better analysis of the cultural practices governing rice cultivation than the KSM project paper, particularly those which affect farm productivity, adversely e.g. lack of preventive measures regarding pests, insufficient or improper combinations of fertilizers, inability to meet irrigation schedules.

The other technical data include the availability of water resources (rainfall, swamps, rivers, reservoirs), water demand and availability analysis and water quality as the basis of computing the demand and supply condition under the project. The KSM project uses a simulation model to project the supply and demand for water to test three options regarding reservoir/pump supply operations using a 13-year data series for rainfall reservoir and water extracted from rivers. In addition, other technical data on soils, topography, drainage characteristics, etc. are provided to facilitate the planning process.

Before calculating the IRR and C/B ratio of the projects, both NWS and NSI selected a sample of farms to estimate farm budgets. In the NWS project, two farm models were used to represent the different production systems obtaining in the area. In Sekinchan, higher yields and better cultural practices contrast with the non-Sekinchan areas characterized by highly variable timing of the arrival and quantity of water which has resulted in the use of low-risk, low-yield techniques of production. Estimates were made (by interviews and consultation) of labour and other inputs used and on the basis of current best practices and MARDI experimental data, future patterns of input use were projected. In order to estimate yields and production, two situations were anticipated in which only the technical irrigation package was linked to productivity gains and where the integrated package's impact (including the agricultural services component) was realized. On the basis of improvements in physical infrastructure alone, cropping intensity was estimated to rise at least to 190 per cent and yield 0.5 ton/ha on various grounds (use of standard wet seedbed, better individual drainage and possibility of burning straw more effectively, optimum timing of maturation and harvesting with periods of high radiation, better control over water to facilitate weed control tillering, and increased fertilizer and insecticide use). Under the impact of the full project package, cropping intensity would rise to 190 per cent and incremental yields of 1.0 ton/ha could be expected. Given the Sekinchan and non-Sekinchan production systems, potential yield and production increases were anticipated to be smaller in the former than in the latter.

In the case of NSI, because of the wide variability in the environment of the 195 sub-projects, three sample schemes were evaluated in detail using DID designs and cost estimates, socio-economic and agro-economic data from the Economics and Statistics Branch of the MOA and field visits. Three farm models in each of the scheme areas were then constructed to project cost and incremental net farm benefits, which were then aggregated to derive, as in the NWS project, sub-project increases in yields and incomes.

The KSM scheme uses the same method except that the data drawn from earlier socio-economic surveys are used to derive average farm costs and returns and projected benefits formulated. Two sets of data were computed for the schemes' sub-project areas, viz. Krian and Sungei Manik.

The IRR and C/B ratio are then calculated and on these bases, recommended for financing. Secondary benefits, e.g. in tree-crop production in the case of NWS and KSM, are also included in C/B calculations. Sensitivity analysis is also applied to determine the impact of changes on capital costs on IRR for the project. In the case of NSI, the minimum acceptable IRR was 10 per cent, although schemes with IRR of 8-10 per cent would be acceptable if at least 50 per cent of the beneficiaries received incomes below the national poverty line and no alternative sources of employment existed in the area.

Project Implementation, Monitoring and Evaluation

Given the integrated nature of the agricultural development projects, there is a conscious need to design implementation and post-project administration structures to ensure that farmers are optionally serviced and income/productivity goals realized. It is of vital significance that the planning environments obtaining in the three project areas do not facilitate a high degree of interagency co-ordination and integration at the farm level, despite the fact that they are irrigated areas requiring effective complementation with agricultural support services. Planning and execution of agency activities are undertaken usually independently, although the State Agricultural Research Development and Extension Committee exists for high level agency interaction. Such a situation persists despite the fact that most agencies come under the Ministry of Agriculture, that the District Action Committees headed by District Officers are meant to co-ordinate district development projects and the FDCs and Village Development Committee provide grass-roots contacts.

In order to ensure that the NWS project implementation proceeds smoothly, a full-time project manager has been appointed, to be guided in all policy matters by a steering committee chaired by the Secretary-General of MOA and the State Secretary of Selangor and consisting of federal and state officials nominated by them. The involvement of the State Secretary will ensure participation of state-level agencies and offices and District Officers to ensure effective programme implementation at the field level. The main responsibilities of the project manager will include : assisting the project agencies in the planning of joint field extension programmes, and in preparing requests for the staff and operating budgets required to implement those programmes; assisting in planning construction of facilities and equipment procurement programmes which will fulfil the various agency requirements; representing the project in periodic meetings with the Public Services Department, the Treasury, and Selangor State government agencies; monitoring and evaluating the progress of implementation actions and project effects and benefits; and reporting on the progress of the project (particularly on behalf of DOA, FOA and BPM) to the steering committee. The project manager, being responsible for the dissemination of new, location-specific varieties to farmers, will have to be a member of the Agriculture Research Development and Extension Committee (on which MARDI is represented). Also, in order to ensure the availability of credit, he will have to project farmers' credit requirements with the assistance of the executing agencies, with BPM having to supply the needs. Given Malaysia's experience that technical officers can perform this management function effectively, a senior agricultural officer has been appointed project manager. The project manager will eventually reside in the area (currently offices are being constructed in Kuala Selangor, the major town in the project area).

Each agency (DID, FOA, BPM, DOA) will be responsible for implementing its own activities specified in the project plan and will appoint its own agency project supervisor who will revert to their departments on project completion. Monthly meetings of agency supervisors will be held under the project manager to review progress, solve problems and plan the coming month's work programme. Every six months a periodic review and appraisal will be made to feed into the six-monthly meeting of the steering committee. It is intended that in the post-project phase, chairmanship of the monthly and six-monthly meetings will be rotated among the agencies under the direction of the State Agricultural Research Development and Extension Committee.

To complement the ministerial and project level of interagency co-ordination, a district level agency co-ordinator will, with his area supervisors, meet with the District Officer monthly to review and assess progress and plan the next month's activities. Like the project committee, the district committee will conduct six-monthly reviews programmed to feed into the similar upper-level exercise. The district level agency supervisors will oversee field operations undertaken through 10 FDCs serving as the mechanism for interagency co-ordination of DOA, DID, FOA and BPM activities. Five FDCs (each under an Agricultural Assistant) will constitute the district subdivisions, co-ordinating subdistrict activities and the work of the field staff operating through the remaining FDCs. These will be headed by an Agricultural Technician, each allocated 16 farmer groups of 50 farmers each and co-ordinating a team of DOA, DID, FOA and BPM field staff making bi-weekly visits with farmer groups. These meetings are to determine, with the farmers, programme priorities at the beginning of each crop season, and provide extension marketing, input supply, credit, water management and O and M services.

Thus an extensive structure is planned with co-ordination among agencies fostered mainly by DOA staff from the project level down to the farmers. The suggested administrative structure takes into consideration the existing structure as, for example, the district office and FDCs. The system of multi-tier operations and reporting upwards will, in theory, provide effective implementation although the crux of the operations hinges upon the degree of interagency co-ordination and integration achieved. The role assigned to farmers provides for a degree of local participation in the project plan implementtation.

Farmers' groups are to be based on irrigation service subareas determined by DID. These farmer groups will be developed by DOA assisted by DID and FOA. The choice of irrigation sub-areas will facilitate farmer group action on water distribution, maintenance and group planting corresponding to the scheduled water delivery dates. As each group is serviced by a team from DOA, FOA, DID and BPM under the co-ordinative management of an Agricultural Technician, similarly a group leader will be elected to supervise the group's activities for extension, irrigation, agri-business and credit, with two assistants to help him. Subgroup leaders would facilitate group activities, each responsible for 10-14 farmers working a 16 ha section. Candidates would be nominated by the farmer and assisted by field staff and *kampong* leaders. Naturally, farmers' group leaders should be *bona fide* farmers working in a designated section. A similar structure for the tree crop areas is to be implemented.

Given the basic framework for multilevel and interagency co-ordination and integration, the role of the field extension officers of DOA, DID, FOA and BPM are also spelled out, to plan for their inputs to farmers. The DOA Agricultural Technician will have a fixed schedule of field visits and cover two farmers' groups a day. His supervisor, the Agriculture Assistant, will visit his subordinates once or twice a fortnight and maintain close contact with the Irrigation Inspector, FOA, area supervisor and BPM credit officer. The district agricultural officer will concentrate on extension but also ensure that regulatory and subsidy programmes fit into the project programme. At the project level, the agricultural officer is responsible for all extension, technical and regulatory programmes, maintaining continuous contact with his line subordinates, scheduling the work programmes of district agricultural officers, assistants and technicians and maintaining and co-ordinating relations with other agencies involved in the project. As supporting inputs, the federal and state agriculture officers provide the extension programme for the project area and arrange with subject-matter specialists demonstration plots, seed production and farmer training programmes and thus assist the agriculture assistants and agricultural technicians in their field operations. Futhermore, the Farm Mechanization Unit of DOA and MARDI provides inputs to train farmers in operation and maintenance of two-wheel tractors and conducts adaptive field research in connexion with the demonstration plot programme conducted by the DOA extension services.

DID will establish a similar structure to that of DOA, providing technical information on water-supply and drainage matters throughout the cropping season, best methods of water distribution within farmers' groups, construction and maintenance of field bunds and on-farm water control structures. Apart from this, DID, in order to protect the interests of farmers' groups within the area, will establish irrigation and drainage committees comprising 16 farmers' group leaders in a service area, the agricultural technician and village representatives. These committees will discuss problems of water distribution and drainage, breaches of the Irrigation Ordinance and obtain local government support for enforcement of rules and operational schedules. Irrigation Division committees covering a command area of 6,000ha will determine the optimum irrigation and cropping schedules desired by each area for the coming season. The district officers, will chair the Irrigation Division committees. At the highest level, an irrigation area committee for the project area will be formed to channel grass-roots requirements up to project level to determine the best irrigation schedules.

FOA is to adopt a similar system in co-operation with DOA and BPM and operate on the same schedules as the other agency field officers. BPM and FOA will assess the credit ratings of farmers on a regular basis and those with a good credit rating will be issued a credit book. This book will contain the farmer's credit requirements (for cultivation, fertilizer, chemicals and labour inputs) as assessed by the group and agency field staff. If approved, a standard agreement requiring the signatures of the group leader and subgroup leader in charge of credit and agri-business will be provided, specifying conditions and lists of approved stockists and locations at which produce must be delivered for loan repayment. To ensure repayment, incentive commissions on loan repayments are suggested for subgroup leaders. For poor credit risk farmers, a system of cross-guaranteeing (using a minimum of three farmers) is suggested. These groups will be given more attention by FOA and BPM.

Finally, all FDCs will be supervised by FOA and operate as centres for input supply, collection of produce and provision of services. LPN will register these centres as official *padi* collection agents. At the subdistrict, district and project levels, FOA and BPM will maintain lists of stockists of inputs recommended by DOA.

Given the unique O and M system recommended, the project plan calls for DOA to chair a working group to formulate management procedures in detail. A pilot project in the Sawah Sempadan area is the basis of the O and M system suggested for the NWS project.

For NSI, containing 195 different sub-projects, the plan calls for a project co-ordinator from DID to manage the project with the assistance of a small staff. His responsibilities would include programming and supervising sub-project preparation, evaluation and implementation as well as over-all administration of budgeting, procurement, disbursement, record-keeping and reporting.

A preparation and evaluation team will also be formed to execute the sub-project preparation and evaluation procedures and programme and monitor agricultural services (extension, input supply, marketing and production credit), and to project charges and taxes and the socio-economic impact of the schemes. The team, to be responsible to the project co-ordinator, will consist of an agricultural economist from MOA, a soil scientist from the DOA Soil Branch, a crop agronomist from the Crop Production Branch of DOA and a DID senior planning engineer. Using a recommended evaluation format (initial review of identified schemes, preparation of scheme proposals, review of scheme proposals, field visits, submission of full evaluation reports, monitoring progress), the preparation and evaluation team, which will maintain professional links to its members' agencies while being administratively responsible to the project co-ordinator, will prepare the agricultural services component of each sub-project for consideration by the Agricultural Services Co-ordinating Committee. This Committee will consist of the Principal Assistant Secretary for Special Projects of the MOA Planning and Development Division, the Director of the Soil and Crop Branches of DOA and representatives of the Directors-General of DOA, MARDI and FOA. The Committee chairman will be the Director-General of DID. The Committee will coordinate the supply of agricultural services and be responsible for programming and monitoring them. Such services include extension, input supply, marketing and production credit for farmers in areas where sub-project irrigation schemes have been completed.

At the state level, the State DID Director will be responsible for pre-project studies and the implementation of irrigation projects. To forge co-ordination, especially to ensure that post-project inputs are effectively co-ordinated by the various agencies, the State Action Committee will be used to bring together the State DID, FOA and MOA Directors, BPM manager, SPU and State Development Officer.

For the KSM project, a less refined structure (compared to NWS) has been formulated for its management and implementation. A project manager will be in over-all charge to carry out the responsibilities of administration, budgeting, procurement, disbursement, monitoring, evaluation and programming of implementation activities by the executing agencies. To facilitate multi-level interagency co-ordination, the project manager will be guided on policy matters by a steering committee with a similar structure to that in NWS, and also participate actively in the Co-ordinating Committee of MOA at state level, the

District Action Committee and State Agricultural Research Development and Extension Committee.

The various agencies involved include DID, DOA, FOA, BPM and MARDI. DID will be responsible for the design and supervision of construction works and subsequent operation and maintenance of the irrigation systems at the field level. DID staff will operate out of FDCs and maintain close contact with farmers through the formation of irrigation service units. DOA will operate as in the NWS project under a project agricultural officer and assisted by inputs from the state extension service. FOA will expand and improve its FDC facilities under the direction of the FOA State Director, while BPM will channel credit via local credit centres (i.e. AFOs) for *padi* production, other crops, agricultural equipment, land purchase, livestock, fisheries and agri-business. The agricultural research component is more extensively formulated by decentralizing the services of the Bumbong Lima Rice Research Branch. To obtain location-specific *padi* strains, research activities will include local varietal trials of rice and other crops; determination of optimum fertilizer input-output relationship; monitoring pest and disease patterns; adapting farm machinery for local use; and assessing the socio-economic impact of research activities. A research team is to be created to carry out these functions and will comprise a project development officer, an agronomist, a plant breeder, a soil scientist and an economist. Besides all these agency programmes, the project manager will supervise and co-ordinate the ongoing programmes (fisheries development, rubber replanting) affecting the agricultural development of the project area.

Linkages with National Strategies

All the projects are defined within the framework of Malaysia's new economic policy of eradicating poverty prevalent in the agricultural small-holding sector. In the NWS case, it has been admitted that the project area is "not among the poorest *padi* or rainfed areas in the country, but in no sense can it be considered wealthy or prosperous". Given that the non-Sekinchan farmers earn on an average $US 235 *per capita* compared with the estimated national relative poverty line of $US 340 and the estimated national absolute poverty line of $US 215, while the national *per capita* income is $US 860, justification for the project is well in line with national poverty redressal strategies. Incomes in the tree-crop areas are even lower, averaging $US 150. Thus the project can be expected to lift the bulk of the poor out of absolute poverty, while those outside absolute poverty will be placed more securely outside. For the NSI project, the identified beneficiaries are among the poorest among Malaysia's *padi* farmers even when double cropping in small-scale irrigation schemes is available and less so when only one rice crop is harvested. While the project is expected to raise real net *padi* farm incomes *per capita* by only $US 54 by 1988 (assuming present farm-gate prices) and raise only 20 per cent of the beneficiaries above the absolute poverty line, it will provide large proportional increases in income to one of the largest identifiable poverty groups in Malaysian society, and provide a sufficient period of time for the planning and initiation of land settlement, industrialization, and other development programmes to assist the rural poor. The same logic for justifying irrigation and other improvements in the KSM project is used as for NWS.

In addition, NSI stresses that the project will reduce urban-rural and ethnic income disparities, reduce national dependence on imported food grains (i.e. make the country more self-sufficient in rice), and provide a stable agricultural base for the peoples of Sabah and Sarawak.

Assessment

On the basis of the three agricultural development plans reviewed and the preliminary observations made earlier on, an assessment can be made of the planning processes involved in the integrated rural development of local areas in Malaysia. This is undertaken to identify areas within which the planning of processes require re-examination or improvement.

Data Base

The data base supporting the technical component of the plans examined is extensive and competently collated, in so far as it strikes the untrained eye of this writer. Water availability and projected irrigation demand are based on historical data series, while the present status of irrigation and drainage infrastructure are objectively and technically appraised. The long experience of DID in identifying and formulating technical feasibility studies in rural Malaysia is reflected in its project plans.

Qualitative data on agency management and performance (DOA, FOA, BPM, MARDI) are collated on the basis of subjective appraisals by the agency itself and the appraisal team. This process, which is quite crucial in optimizing the productivity of project works (like drainage and irrigation infrastructure), requires a more systematic study more objectively to delineate functions and management performance as a basis of projecting duties and responsibilities required under the project plans.

The quality of socio-economic data used reflects two deficiencies which can be quite serious in their effect on project estimations and projections. One weakness relates to the quality of socio-economic data dealing with the farmer and his social environment. Where sample surveys are undertaken, there tends to be insufficient coverage of farm size, tenure status and relations, current cultural practices, debts, organisations and institutions and, particularly, farmers' attitudes, and behaviour with respect to new technologies and change. Connected to this is the tendency to treat the farming community as a homogenous group, rather than to focus studies and project attention (of the agencies responsible for the implementation of the agricultural components, for instance) on the poorer, underprivileged strata of rural farming communities. Also, a study of historical trends in farming (e.g. farm size, rentals, mechanization) could generate useful information in formulating activities to support or offset developments as part of the agricultural development planned for the project area. The second deficiency pertains to the estimation of incomes, yields and production increases by using micro-level studies and experimental data. The problem then arises of aggregating the farm behaviour of thousands of farmers working under different agronomic and other conditions. In NWS, two farm models are used to represent 32,000 farmers. In NSI, three farm models are projected to reflect the behaviour of 60,000 farmers operating in 195 different small-scale irrigation schemes scattered in West Malaysia, Sabah and Sarawak. The KSM project utilizes a small sample study of rice farmers in the sub-project areas and averages out data for current costs and returns to farming. The small size of the sample (96 farmers in the Krian irrigation scheme) is restricted, we are informed, by convenience and a lack of resources. Given the current costs and returns to farming, projections of yield increases (under various assumptions) are normally made by using a factor equivalent to the best cultural practices obtaining in an area and experimental data on new varieties to be used in the project. Such a data base is used to project the major expected benefits from the agricultural development of an area based on technical improvements and more efficient extension of the agricultural services component and as the basis of investment decisions in the range of $US 60-90 million.

Project Management

All three projects call for the appointment of a project co-ordinator or project manager, usually a senior agricultural officer or DID engineer, to be in over-all charge of project implementation. Assuming that multi-level co-ordination among the executing agencies can be successfully attained, on project completion, it is assumed that the project manager or co-ordinator will continue to operate. In the NWS project, the project management will be rotated among the executing agencies under the direction of the State Agricultural Research Development and Extension Committee. No mention is made of the co-ordinator in NSI or KSM. Whether, on project completion, the management system can continue to function effectively in delivery agency inputs to the beneficiaries remains to be seen. Past experience in rural Malaysia of similar projects is ambivalent: project managers in the post-implementation phase of an agricultural development scheme have been effective in co-ordinating services (MADA) or have met with various obstacles (KADA and Besut). The fact that the project planners are unwilling to suggest the adoption of the MADA management structure, in which the general manager exerts strong leadership in the integrated agricultural development of the area, indicates that the planning environment does not support the recommendation of such a structure. Or it may be that the planners leave it to later initiatives and effort that such a structure can be evolved.

The management structure formulated projects for the

attainment of interagency co-ordination by a system of continuous monitoring and supervision of lower levels of the management hierarchy by upper-level committees. Although no explicit statement is offered, one suspects that failure to obtain interagency co-ordination at any particular level can be corrected through direction by the agency supervisor or committee at the level immediately above (Agricultural Technician—Agriculture Assistant—District Agricultural Officer—Project Manager—Director-General/State Secretary, in the case of NWS). Co-ordinative strategies at any one level are not explicitly spelt out; only the identification of the chairman of the committee of agency officials operating at different levels of the project structure is made. An input by administrative experts could be useful to evaluate current bureaucratic and administrative structures and the new forms recommended under the project.

The formulation of the management structure of the project does not fully integrate into the existing bureaucracy or organizations in a project area, either at the planning or execution stage. In all three projects, the role of the district officer, who co-ordinates interagency activities in the district and supervises project implementation of the district development plan, is taken into consideration not at all or only marginally. Neither is any mention made of village development committees or the Community Development Branch of MOA (KEMAS), especially important if the ideology of popular participation is to be put into use. Similarly, the existing institutional structure of rural co-operatives, farmers' associations and small agricultural units, although these form an integral part of the area farmers' organization, is not related to the proposed project management structure. An evaluation of the administrative and organizational apparatus existing in an area appears necessary, particularly in view of the need to prevent the creation of an even more cumbersome administration than before or isolation of existing bureaucracies and organizations from the new project administration.

Institutional Development

Technical competence in the design and implementation of drainage and irrigation infrastructures, for instance, does not appear to be lacking in agricultural development planning. Planning inputs in institutional development or institution building, particularly of farmers' organizations and farmers' groups, does not have depth of experience or refinement compared with technical developments. This must be a crucial area in agricultural development planning as the focus of the project on farmer beneficiaries and the delivery of services and inputs are eventually facilitated through groups or institutions representing farmers. What also appears to be overlooked is that the structure of farmers' groups must be such that the delivery of project inputs and services must be optimum; delivery systems must interface efficiently with the structure proposed.

The basis of institutional development in NWS is a group of 50 farmers having a common reference to a small irrigation sub-unit. While this would appear to be a rational basis for creating groups (eventually merging into irrigation committees, division and representation at the project level), more research into farmers' attitudes to such groups, choice of leaders, effect on non-formal groupings in villages, the role of the *sidang* or sub-leader appears to be necessary. The success of such traditional groups in Indonesia or the workings of experimental groups in Sawah Sempadan is no guarantee of success or complacence among planners. Further, if the farmers' groups and irrigation committees are to be the basis of future efforts to create *bona fide* rural institutions, this should be an area of intensive research in the project plan, undertaken with the co-operation of all implementing agencies. The future of the Farmers' Organization Authority, which was created to integrate farmers' co-operatives and farmers' associations and motivate farmers into forming self-reliant and dynamic communities, will also have to be considered and evaluated.

In the area of institutional development, it can be observed that local area plans for integrated agricultural development do not seem to have any mechanisms whereby more local institutions (district officers, for instance) or the grass-roots (village development committees, farmers as a whole) can be consulted prior to project formulation. This is in part due to the large area under the integrated agricultural development plans studied. It would appear that a systematic programme of consultation and participation can be made by disaggregating the planning process into districts and institutions below. In other words, the basis of local area plans (covering 30,000-50,000 beneficiaries) should be a district integrated agricultural or rural development plan which utilizes socio-economic profiles and consultation with people in order to define and formulate district-wide development plans that eventually merge into regional, state and national plans. Such a system of "bottom-up" planning will complement the "top-down" approach that currently utilized more because it is a simpler and quicker planning process than because of ignorance of the need for grass-roots inputs into the planning process.

Research and Evaluation

In each of the project schemes, a conspicuously missing section is a research and evaluation division in order to provide informational feedback not so much on project implementation schedules as to generate important data as a basis of systematic and continuous appraisals of the technical social and economic impact of improvements in an area. Such an activity is required not only to measure the attainment of project goals (increasing incomes and production) but to provide early warning signals of impacts or trends that could not have been anticipated by planners. A case in point is MADA which recognized the need for socio-economic evaluations of agricultural development efforts only some time after project construction began. It now has a fully-fledged economic and evaluation division providing analyses which has become a definite input in the management of MADA. Some of the trends that could be analysed are the effects of mechanization in improved double cropping areas, patterns of labour utilization and supplies in *padi* and non-*padi* activities, changing attitudes and norms of farmers, disintegration of traditional institutions, the need for strategies to absorb surplus labour, etc. Furthermore, the basis of further developments in an area will have to be based on data and analysis generated by the research and evaluation division of the project itself.

Training Inputs

A notable feature of the NSW and KSM projects is the recognition that training inputs are required to enhance the skills of both field personnel and farmers. In the same way, there is a definite need for improving the skills of planners involved in agriculture and rural development programming to improve agency plans, facilitate integration of agency plans into a project or master plan and finally into a state or national development plan. Agencies which could benefit immensely by training in planning processes are the Ministry of Agriculture Planning and Evaluation Division, state Planning Units and state or regional branches of the Economic Planning Units. The inputs of some of these agencies in local area plans are minimal, but their impact can be beneficial in assisting agricultural development agencies to improve their planning skills and district officers in producing district development plans.

Incentive Policies in Planning

It is a truism that a built-in deficiency in agricultural planning is that plan implementation hinges upon the decision of thousands of farmers who may or may not behave as planned or anticipated. Thus, for example, IRRs are projected on the basis of farm community acceptance of modern technologies in an optimum package. Where previously Governments have depended on direction to farmers in order to influence acceptance of new technologies, training of farmers has now become an important policy instrument to disseminate technology and influence its adoption on the basis of learning situations through experimental plots, etc. A complementary policy strategy that could be utilized in plans would be the use of economic incentives to make it profitable for farmers to adopt a modern technology package via input subsidization policies or stable product price strategies. All the plans assume the availability of inputs such as fertilizers, chemical insecticides, labour or machine services when farmers need them, but price relations are not considered in determining or influencing the demand for such services. Similarly, disincentive (tax) policies may be used to discourage the cultivation of crops that are not agronomically suitable.

In sum, the "stubborn" peasant requires not only to be trained to be a modern farmer ; he must be economically induced to exhibit behavioural patterns that are economically desired and rational. In the words of Dandekar :[1]

Let us then ask : what can the State do in respect of items

in which the ultimate decisions lie with the farmers? As I see it, three things : (1) educate and improve the farmer as a farmer ; (2) reorganize the production apparatus in agriculture so as to enable the farmer to take better care of his land and water resources ; and (3) create appropriate institutions in order to improve the decision-making in agriculture.

The first is a task of education. The second is a task of much detailed work on the ground. The third is a political task. We have neglected all three because of a mistaken belief that we could achieve production targets directly without bothering to improve the man to improve the land, and to improve the institutions governing the relation between man and land.

Integrated Development : Agricultural or Rural

Integrated agricultural development planning aims at clearly defined income or productivity goals and also implies a greater potential for employment generation in a locality. However, there is a need also to consider within the process of agricultural planning, the goals of equity, access to resources and even control over resources by the beneficiaries of development plans. This involves not only a greater planning input in institutional development but also a meshing of plans related to such services also focused on the farming community ; local government, hospitals, electricity, water, schools etc. It is not being suggested that project plans like NWS be broadened to include these items in a strategy for integrated rural development, as the project manager could better and should devote his talent to obtaining interagency co-ordination and supervising scheduling, monitoring and implementing the project plan. However, agricultural planning should gradually phase in the involvement of other development agencies and, particularly, the district officer and farmers' institutions so that the income and productivity gains consequent upon agricultural development serve as a stepping stone to the broader development of the area.

Integration with National Economic and Agricultural Policies

Although admittedly this is an objective or goal that can only be realized when planning skills and techniques are refined, there is a need to plan for the integration of local area and district plan with sectoral and national plans for agriculture, rural and national development. Linkage between an area and other parts of a region or State need to be rationally developed to allow, for instance, for surplus labour resources to be channelled to land development schemes, or for excess saving owing to rising incomes to be mopped up by taxation or other measures for agro-industrial or other development projects. A first step would be the formulation of a national policy on agriculture so as to ensure consistency in the various sections of the agricultural economy. In the case of Malaysia, the setting up of a national working committee to formulate such a policy is a tangible move towards this goal.

REFERENCE

1. Quoted in W.W. Cochrane, *Agriculture Development Planning Economic Concept, Administrative Procedure and Political Process,* (New York, Praeger Publishers, 1974), p. 98.

FOURTEEN

NEPAL

Bihari K. Shrestha

Past Experience : A Summary*

NEPAL is a land-locked country with a population of about 12 million, of which 60 per cent live in the hills and mountain areas, and about 40 per cent in the plains bordering India (*terai*). The country is subdivided into 75 districts that are regrouped into four broad development regions.

During the 1950s and 1960s, rural development programmes were initiated in several districts under the guidance of United States, Indian and Swiss advisers, among others. In 1960, the King abolished parliamentary democracy and instituted a partyless *Panchayat* system at the village and district levels for the purpose of local administration and development. *Panchayat* Development Workers (PDW) were assigned to work with the village *Panchayats* and help them with plan formulation and implementation. At the district level, *Panchayat* Development Officers (PDO) served as executive secretaries for district *Panchayats* and were assisted by a team of technicians in the fields of engineering, agriculture, home science, co-operatives etc. under the guidance of their respective ministries. Most of the planning was actually done on a sectoral basis by the departments concerned, and the role of the PDO gradually eroded. Special guidelines were issued in 1965, on the decentralization of government functions, but local officers were reluctant to submit to the control of popularly elected district *Panchayats* and the ministries concerned were equally reluctant to lose their own control, which tended to defeat the original purpose.

* Summary prepared by the ESCAP secretariat.

Current Approaches to Local-level Planning for Rural Development

There are currently several different rural development programmes being implemented in the country which carry different implications for a design of rural development in general and local-level planning in particular. Some of them are reviewed below.

District Administration Plan

In order to find a workable and efficient solution to the problems of development administration at the district level, the District Administration Plan was drawn up in 1974 based on the premise that it takes an integrated structure to provide the needed services and facilities to the people in an organized and co-ordinated manner. The Plan stipulated that the district administration would be a unified entity which would streamline the role of the district and village *Panchayats* and would work out a stable and rational basis for interagency co-ordination. To that end, the Plan provided for the following :

(i) All district-level agencies of the Government were designated as the constituent sections of a unified district office;

(ii) Planning for sectoral developments was to be initiated at the village level and an integrated document called the District Development Plan was to be prepared annually with the implication that the nationwide programmes of individual departments would, to a considerable extent, become an aggregation of district sectoral plans;

(iii) Several subject-matter committees (called co-ordinating committees) were formed of the district *Panchayat* members and concerned technical officials to enlist popular participation in decision-making for all developmental programmes in the district, thus bringing popular institutions within the fold of a unified district administration;

(iv) The Chief District Officer (CDO) would play the role of the co-ordinator of all development programmes at the district level, act as supervisor of the district personnel, and function as the chief implementer of the district *Panchayat* decisions. Briefly, while the district *Panchayat* would play a significant role in voicing the wishes of the people in plan formulation and resource allocation, the CDO would represent the focal point of a unified district administration.

However, even after many years of passage of the Plan, it is still not fully operational. While the semblance of a unified district administration has been achieved by renaming the district level agencies sections of a so-called District Office, the departments and ministries concerned have for all practical purposes refused to entrust the supervisory and co-ordinating role over their programmes and personnel at the district level to the Chief District Officer, thus affecting one of the most critical provisions of the Plan.

As it stands, planning in most of the sectors is done separately at the national level. These nationally-set targets are later disaggregated and handed down to individual districts, and it is these targets that are given the seal of approval by the aforesaid co-ordinating committees and incorporated into the so-called district development plan. This is later presented to the district *Panchayat* and the district assembly for their approval, which they simply cannot withhold. In sum, what was supposed to be a genuine planning exercise at the district level turned out to be a programme of implementation of the plans and targets formulated at the national level. Planning of programmes on the basis of local identification of needs and potentialities essentially remained undone.

There none the less exists an interdepartmental committee at the national level to implement the District Administration Plan which was originally prepared by a special task force and later approved by the Government with dispatch. But with the passage of time the ministries discerned provisions for effecting major deconcentration and devolution of their powers and functions, and held their ground even more steadfastly. No substantial attempt has been made to overcome this problem and to put into effect the many useful provisions of this novel endeavour in local development administration.

Because of the sectoral approach to local development planning, progress monitoring is also done at the sectoral level. Reviews and evaluations of the programmes at the district and village level with a view to ascertaining their impact as distinct from the progress of target accomplishment, are rarely done. Such an evaluative effort has been an equally rare phenomenon even at the national level. Progress reports are essentially reports on the accomplishment of targets originally set at the centre and handed down to the local level. For want of an effective supervision mechanism, tampering of progress reports to "meet" the original target is also frequently resorted to.

Panchayat Sector Programmes

The national planning of Nepal provides, in addition to the two conventional public and private sectors, a third sector called the *Panchayat* sector. It is based on the premise that by making use of the longstanding tradition of local self-help in the society additional resources can be generated, although in the form of voluntary labour mostly, for local development efforts which could go a long way in maximizing development investment in the country. For this purpose the fifth Five-Year Plan has provided for an investment programme worth 253 million rupees under its maximum programme[1] for *Panchayat* Sector activities which are expected to generate an extra resource of 934 million rupees locally. The sector comprises the following programmes.

LOCAL DEVELOPMENT DEPARTMENT PROJECTS

The first category of *Panchayat* sector activities comprises the projects undertaken by the Local Development Department (LDD) of the Ministry of Home-*Panchayat* which consist of small, local self-help schemes such as drinking-water projects, footbridges, small irrigation channels, school and *Panchayat* buildings, tracks and trails, etc. These projects are undertaken by the district and village *Panchayats* with grants made available by the Department and they are responsible for eliciting matching contributions of voluntary labour, money and materials from the villagers according to the proportional standard set by the Department on the basis of the nature of the project and the resource endowment of a given geographical area.

The planning process adopted for this programme provides for more decentralized decision-making than other rural development programmes, such as agriculture, industries etc. Resources are estimated at the national level on the basis of past years' experiences as well as the estimated availability from donor agencies such as the United Nations Children's Fund (UNICEF) which provides assistance to LDD for rural drinking-water schemes and roofing materials for schools.

Most of the resources are disaggregated and sent to different districts, the quantum being determined mostly on the basis of population of the districts. Where bigger schemes are involved, such as suspension bridges or UNICEF-assisted drinking-water projects, the assistance in kind is mostly emphasized because of the inability of the districts to procure external materials such as cables or polythene pipes and to ship them to the project sites which most often involves shipment by air, including occasional helicopter ferrying.

The selection process for all the projects is uniform. Projects are first identified at the village level, passed by the village *Panchayat* first and then by the village assembly. Where external assistance is required, they are sent to the district *Panchayat* which draws up its own priority and criteria, and allocates the grants to different village and district-level projects (the latter consisting of the relatively bigger ones) accordingly.

The district *Panchayat* also provides technical support to the projects through the District Office. Where the technical capability at the district level for a given project is inadequate, the regional offices of the Department provide it. This is where the role of a fairly large contingent of volunteers from the Federal Republic of Germany, Switzerland and the United States becomes important.

REMOTE AREA DEVELOPMENT PROGRAMME

In order to provide additional impetus to the development of basic services in the remote (mostly northern) areas of the country, there exists a special committee, called the Remote Area Development Committee, within the framework of LDD which includes non-official members from those areas. The Committee, often by personal visits by its members, solicits projects from those areas and allocates its resources on the basis of specific consideration for specific districts and projects. These decisions are then handed down to LDD which implements them through the respective district and village *Panchayats*.

HILL TRANSPORT DEVELOPMENT PROJECT

The World Food Programme has for a few years made assistance available to Nepal under its Food for Work Programme. This assistance consists of such edible commodities as powdered milk, flour, and oil. The Government has constituted a committee, also within the framework of LDD, to construct major hill tracks and trails according to the demands of the local people. Payment to the workers is made in kind, each worker getting enough for a family of five. No voluntary labour is stipulated under this programme. Although the District Office is the mainstay for project implementation, the decision to construct a certain hill track or trail is taken at the level of the committee and is handed down to LDD for execution.

REGIONAL INTENSIVE DEVELOPMENT PROGRAMME

An important project under the *Panchayat* sector is the Regional Intensive Development Programme which was until recently known as the Small Area Development Programme (SADP). The incorporation of the programme in the current Five Year Plan of Nepal symbolizes the increasing importance attached to the concept of regional planning introduced in the fourth Five-Year Plan. With growing experience in national development it was increasingly realized that there existed an immense regional disparity in the resource endowment and distribution of developmental benefits leading to unprecedented movements of people from the hill and mountain areas to the *terai*. Therefore, in order to offset these unbalanced growth trends, the concept of regional planning was introduced which meant the development of the hilly areas of Nepal in complementarity with its *terai* areas, through judicious allocation of scarce resources in selected areas with potentiality for growth, and interregional circulation of goods, services and people. For this purpose, four north-south "growth axes" or development corridors were identified at approximately 80 to 100 miles distance between one another in the country. and along each corridor "growth centres" were identified which were to be linked with a road artery (either stipulated or already in existence) and which were to function as polyfunctional settlements to cater to the diverse needs of the hinter and.

As part of this regional development approach, a programme was conceived under which 20 small areas were identified in the current Five-Year Plan which were located around the development corridors.[2] The selection and delineation of the small areas for operational purposes was on the basis of the following criteria. Firstly, the area should be small enough to be manageable which, in operational terms, meant an area within a day's walk from its centre, with a population of approximately 20,000. Secondly, the area should be endowed with identified potentiality for growth. Thirdly, it should be close to a modern transportation facility such as a road or airfield.

While this approach presupposes resource concentration on a selective basis through a so-called mutually reinforcing integrated programme, the rationale behind it is to make development more visible and more rapid than it has been hitherto. It is further stated that, if the available resources are thinly spread, the opportunity to build up critical masses at a relatively few places and to let them generate self-augmenting externalities will be missed. To that end, the Five-Year Plan envisaged an allocation of 5 million rupees to each small area under its maximum programme.

However, within two years of implementation of SADP, political pressures from the people living in adjoining areas started to mount, and the Government yielded in 1978 by extending the small areas to include the entire districts in which they were located. Keeping in view the consequent need for expanded resources, for that purpose, the Government further decided to limit the programme to those districts where it has been in operation instead of extending it to all 20 places, as stipulated in the Plan, and to change it to the Regional Intensive Development Programme (RIDP).

Despite these changes in the programme with far-reaching implications for its very conceptual basis, the operational objectives and its structure remain generally unchanged. There are fourfold objectives, namely, increase in productivity, increase in employment, increase in the level and coverage of the social services, and development of infrastructure. In order to achieve these objectives a package of programme components has been worked out which include such areas of activity as agriculture, forest development, cottage industry, health, education, nutrition, co-operatives, and a women's programme. As part of the programme a multipurpose training unit to be run by the subject-matter officers of the integrated programme has been stipulated. To ensure regular and timely reviews of progress a reivew unit has also been provided for.

Allocation of resources to these activities under this programme is to be made in addition to the allocation generally made to all the districts by different departments under their sectoral programmes.

The programme is directed by a central RIDP Board under the chairmanship of the Minister for Home-*Panchayat* and the vice-chairmanship of a member of the National Planning Commission. Other members include all the heads of departments and senior officers of the ministries concerned with rural development. Its functions include laying down policies and establishing inter-agency co-ordination for the programme.

At the district level, the programme is designed and implemented according to the provisions of the District Administration Plan described above. It is proposed that at the beginning of every fiscal year every department and ministry involved in the programme will inform the districts concerned of the quantum of resources likely to be made available to them, and on this basis the district *Panchayat* will solicit projects from the village *Panchayats*, process them through the co-ordinating committees, develop an integrated programme, and implement them.

However, as stated above, the individual departments have normal programme activities alongside the RIDP allocations in these districts. Such an arrangement has caused considerable difficulty in eliciting serious interest from the district-level agencies of these departments. While their performance in the implementation of the programmes handed down to them by their central departments is characterized by a generally low level of motivation, they consider the RIDP component an unwelcome addition and therefore optional. Thus, integration and co-ordination at the local level as well as at the central level (despite the Board) have become a major problem of the programme. Co-ordinated training programmes and a joint review system at the local level have not yet come into existence. Where training units have been set up they conduct training programmes in isolation without reference to or support from other subject-matter programmes being implemented in the district.

Despite these built-in weaknesses, however this programme has recorded some relatively noteworthy progress in terms of number of projects undertaken and completed in these remote areas. Most such projects consist of local self-help schemes, such as drinking-water projects, irrigation channels, school buildings, and tracks and trails. The reasons to which this progress can be attributed are the decentralization of the planning function to the village and district *Panchayats*, and the support and follow-up services provided by a special functionary called the SADP Officer acting as the promoter of the programme among the villagers as well as the district-level officials.

RASUWA-NUWAKOT RURAL DEVELOPMENT PROJECT

Like RIDP, RNRDP is another project under the *Panchayat* sector. Rasuwa and Nuwakot are two contiguous mountain and hill districts, situated immediately to the west of Kathmandu, with a combined population of 190,000. Of the two districts, Nuwakot has a road link to Kathmandu which was constructed as part of a hydro-electric project installed there. This place was also the scene of a government/FAO/UNDP multipurpose study project in 1967 and an FAO/IBRD co-operative programme study in 1974, the latter to identify and prepare a project for the development of the two districts. Possibly because of this previous donor interest in the area and certainly because of the benefit of an already existing road link, the two districts were identified as possible places for a regional development project in the Fifth Five-Year Plan. This project has since been undertaken as a five-year Rasuwa Nuwakot Rural Development Project with a World Bank Group loan of $US 8 million and a total outlay of 136 million rupees ($US 1 = 12 Nepali rupees) beginning in 1976/77. The project is based on the appraisal report of a team sponsored by the International Development Association (IDA) in 1976.

The project is intended to support the Government's development strategy which seeks to balance economic growth with income distribution, provide for more equitable regional development and ensure productive benefits from previous road investments. To that end, the project consists of activities in the fields of agronomical, horticultural, livestock and dairy development, hill terrace and river valley irrigation, agricultural inputs and credit, co-operatives, cottage industries, forestry and erosion control, health facilities and drinking-water supply, and track, trail and footbridges.

At the district and village level, the project is implemented according to the provisions of the District Administration Plan. Between the district and the village *Panchayats*, however, there is also a third tier, called the subcentre, of which five are planned for Nuwakot and two for Rasuwa. Project activities are stipulated to be decentralized to the subcentres from where training and extension services are rendered to the village *Panchayats* in their area.

There is also an Office of the Co-ordinator for the project in the Ministry of Home-*Panchayat* which is expected to perform several important functions in relation to the project at the centre as well as in the districts. Such functions include, among others, assisting the district in planning and plan implementation, establishing liaison with line departments, and monitoring the progress and impact of the project. The Co-ordinator is assisted by an administrative and an evaluation expert, both expatriates.

There is a Project Co-ordinating Committee (PCC) under the chairmanship of the Secretary of the Ministry of Home-*Panchayat* on which all concerned line departments are represented. Itsm ain function is to assume over-all responsibility for the co-ordination of project planning and implementation and for that purpose, it is expected to meet fairly frequently.

Planning at the district and village level is to be done according to the procedures laid down in the District Administration Plan. It means that planning in different fields of activities is to be initiated at the district level by the subject-matter committees and approved by the district *Panchayat*. Planning for projects to be financed from local resources with or without government grants is to be initiated by the village *Panchayats* and approved by the village assembly, upon which projects will be presented to the district *Panchayat* and the district assembly for necessary funding.

The District Development Plan which is drawn up annually at the district level is an integrated totality of the sectoral plans put up by the different co-ordinating committees or subject-matter committees. Indications as to the possible amount of resources to be made available in a given year are provided by the line ministries in advance. Once the plan is approved by the district *Panchayat* and the district assembly it is sent to the line ministries for approval and funding.

As stated earlier, DAP requires that the plans drawn up at the district or the village level are based on assessment of the needs of the local people. When prior study of the district and projects recommended therein is available, it is expected that this would be taken into account in the planning exercise at the district level and the village level. However, the fact that such a dynamic view of the process is lacking on the part of the line agencies has already been discussed. In the case of the RNRDP, the line agencies simply copy the figures given in the appraisal report and hand them down to the district level agencies in the two districts for incorporation into their respective district development plans. Even where gross incongruities with changed circumstances have been noted, departure from the appraisal has been infrequent. Besides, lack of genuine planning exercise at the local level has deprived the project of the benefit of intercomponent integration which is most often critical in making a desired impact on the problems of the area.

Progress monitoring is considered to be an integral part of the project and is stipulated to be conducted at two levels, namely, the district under the over-all supervision of the CDO and the centre by the Co-ordinator. The Co-ordinator is also expected to prepare a final monitoring report on a quarterly as well as an annual basis which is presented to the Project Co-ordinating Committee and thereafter to IDA.

The project design also provides for an objective evaluation of its socio-economic impact on the people and for assessing whether the project does provide a model for replication, in other hill areas in the country. It is expected that there will be a baseline study of the project and an ongoing evaluation on the basis of detailed monitoring of selected areas.

In respect of both the activities, i.e. project monitoring and evaluation, they are performed only with partial effectiveness. While progress reports are put out at different intervals as required, they are done more in fulfilment of formal requirements than as an instrument of programme review and performance improvement.

Similar is the case with evaluation. A baseline survey of a sample of project villages has been completed. However, one of the conclusions is that, given the objective of the project to bring the population of the project area to subsistence level, only the small holders (60 per cent of the local population) are its target population, which calls for reshaping of the project design itself. Little has been done in this direction, and the project continues as if it were addressed to an economically homogeneous population.

Panchayat Development and Land Tax (PDLT) Programme

In order to generate resources from agriculture and plough them back into the local economy by investing in the local development programmes, this novel system of taxation was first introduced on an experimental basis in 12 villages *Panchayats* of an eastern *terai* district, Jhapa, between 1966 and 1968. The main features of this tax are the following :

(i) The tax, which is levied in replacement of the regular land revenue, is based on the produce of the land rather than on its area as before, and with increasing productivity this is expected to yield progressively larger amounts of revenue;

(ii) Unlike regular land revenue, this tax is shared by landlords, tenants and ownertillers alike. While the landlords pay 15 per cent of their produce, the tenants pay 3 per cent of their share of produce if it is less than 50 per cent of the produce, 5 per cent if it is 50 per cent or more, and the owner-tillers pay 6 per cent of the produce;

(iii) A major proportion of the proceeds from this tax is expended for local development. Of the total proceeds, 35 per cent goes to the Government in lieu of the land revenue which it refrains from levying in the tax area, 10 per cent to the district *Panchayat* as a contribution to the district development fund, and the rest (55%) is retained by the village *Panchayat* itself to finance its development projects. Hence the name of the tax.

While this is attended by several built-in or circumstantial shortfalls, e.g. lack of audit and possibilities for misuse of funds, difficulty in collecting the tax from tenants (who are not required to share the regular land revenue which is paid only by the landlord) for want of punitive measures against defaulters, its potentiality for exacerbating regional imbalance which already favours the *terai* area, lack of information on its impact on local development as well as on different castes and classes of people, and the apparent proclivity of the village *Panchayats* to allocate these enhanced resources predominantly for social service projects, the fact remains that this experiment has been instrumental in mobilizing, and retaining for local development purposes, significant sums of resources in the selected *Panchayats*.

One important implication of this tax system for local development planning is that it assigns a major role to the village *Panchayat* compared to the roles of the district *Panchayat* or the district office. Under this system the role of the latter is limited to providing such technical and support services which, owing to their limited authority and administrative capabilities, they cannot muster on their own. Projects are mooted by the village *Panchayat* and approved by the village assembly upon which they are then implemented. Reference is made to the district *Panchayat* or to the district office only when technical guidance is required.

Despite the relatively comfortable resource position of the PDLT village *Panchayats*, their capability in the sphere of planning for integrated development of their area is limited for structural reasons. In a situation such as that in Nepal, where development is a function not only of financial resources but also of adequate institutional and policy supports from the Government, the inability of the latter to respond adequately to the expanded needs of the village *Panchayat* has forced them to use their resources less on economic development projects such as agriculture (which require intensive input of services in extension, agricultural inputs, irrigation, credit, marketing etc.) and more on such special service projects as education and drinking water. For instance, in 1970/71 the 12 village *Panchayats* made an investment of 1 million rupees out of their own resources in contrast to 150,000 rupees by a sample of 12 non-PDLT *Panchayats.* However, of this resource, the PDLT *Panchayat* spent 73 per cent on such headings as education, road building, and other construction activities, 26 per cent on irrigation and drinking water, and only 0.8 per cent on the heading called agriculture development.

Lately the Government is taking full stock of the situation and has since extended the system with needed amendments to a few more districts in the country.

Small Farmers' Development Project

Under the joint auspices of the Food and Agriculture Organization of the United Nations (FAO) and the Agricultural Development Bank of Nepal, a small farmers' development project is being implemented on an experimental basis in a few selected village *Panchayats* in a hill and a *terai* district in the country. The distinctive features of this project are:

(i) Formation of groups of small farmers (who are defined as those households owning 0.4 ha or less, including landless, in the hills and 2.6 ha or less in the *terai*) on a voluntary basis;

(ii) Extension of loans by the district agricultural bank through the local co-operative societies for individual and group projects of the members of the groups such as agricultural, horticultural, livestock, or cottage industries projects on the recommendation of the group as against simply individual application;

(iii) Group liability for individual loans thus making an element of mutual trust and group supervision of individual projects an integral part of the arrangement;

(iv) Adoption by the members of other beneficial practices such as building and use of latrines, family planning practices, and so on;

(v) Provision of assistance by a project office located in the *Panchayat* itself in the identification of needs by the group members, processing of loan applications, provision of loans through locally established co-operative and other supports from the district-level offices of the Agricultural Development Bank, health, cottage industries, etc.;

(vi) The formation of a district-level interagency co-ordination board, under the chairmanship of the district agriculture development officer, to provide necessary support to the project.

While the project is still in its experimental stage, results have been generally encouraging, and the number of groups and their projects are steadily increasing. Collection of loans by the local co-operative society has never been so timely and easy. Even those small farmers living outside the project area have picked up the idea and formed groups voluntarily, and persuaded the Agricultural Development Bank to extend them loans on group liability. The groups have even acted as a political force in the village *Panchayat* in the sense that the office of the *Panchayat* president is held by a small farmer who has demonstrated his all-out enthusiasm for the continued expansion of the project.

While this project has not been as uniformly successful in the *terai* as it has been in the hill area, its current success in the latter can be tentatively attributed to a combination of the following factors :

(i) The egalitarian composition of the groups;

(ii) The inherent decentralization of the function of need identification to the people affected by it;

(iii) Decentralization of the supervision function to the people, i.e. the members of the groups, who can perform it more timely and effectively than any outside agency;

(iv) Co-ordinated input of resources for the group and individual projects;

(v) Evaluation of the project performance by the affected people themselves who in this case happen to be the members of the groups undertaking the projects;

(vi) A rational organization for the programme providing support services affording the groups easy and timely access to necessary services.

A National Strategy for Rural Development

While the aforesaid approaches to rural development currently in practice in Nepal demonstrate clearly the unintended heterogeneity, more such projects are already in the pipeline. Preliminary work on a rural development project in parts of the eastern hills are being undertaken with United Kingdom assistance. The Asian Development Bank is finalizing a project for three districts in eastern Nepal. The Canadians are also doing their own preparatory work in the two zones in far western Nepal. And the World Bank is putting together the necessary information for its second project in the hills of the westernmost zone of Nepal.

Clearly, there was a need for a national approach to rural development. In 1978 a strategy for rural development was worked out by the Government which turned out to be mostly based on the design of the IDA-assisted RNRDP and which, it was said, was originally drafted by the World Bank itself as Nepal's strategy for rural development. It was later presented to the meeting of local representatives of the members of Nepal Aid Group.

The recommendations of the strategy were primarily concerned with issues at the national level. For instance, it stated that programmes would have to be resource realistic and should embrace the integrated approach. Projects should be based on road access to facilitate inputs for rural development. They must also have redistributive effect and, to that end, should emphasize labour-intensive projects. Lastly, they should be aimed at the gradual elimination of inter-regional and interpersonal differences.

The strategy continued that in order to achieve those conditions rural development should involve people in project planning and execution, should be based on the needs of the people and should be planned with reference to the micro conditions. People's participation must be ensured in all the planning, programming and implementation phases.

However, when it came to the organizational structure for implementation, the strategy was less than explicit. It stated that "the project/programme design would be the responsibility of the co-ordinating office and the various line departments and the National Planning Commission, the prime responsibility for project implementation rests with the district offices of the participating ministries". And recommended that "annual district planning is initially formulated by each ministry or agency in conformity with the project guidelines and project implementation schedule". It further stated with implicit contradiction to the preceding propositions that "the district secretariat prepares a district development plan in consultation with the district *Panchayat*".

In sum, while the strategy represents a necessary attempt to fill the void in this area, the details set out therein are in many instances at variance with the principles recommended for a rural development programme in Nepal.

Over-all Assessment of the Situation

Since the inception of the *Panchayat* system of democracy in the country, one of the most outstanding achievements has been in the area of institutional penetration of Nepal's entire rural areas by the establishment of village *Panchayats* as the local organization for development and administration. Similarly, the conception and introduction of the District Administration Plan also represents an important national attempt to decentralize the developmental powers and functions of the Government to the district and village levels, and to establish (implicitly if not explicitly), a system of integrated development planning at the district level.

Despite such conducive institutional framework and national endeavours, the approach to rural development in general has differed at different times and under different projects. Because of this confusion, planning at the local level has mostly been dictated by the specific design of specific project, and not based on a premise generally shared by the constituent agencies of a rural development programme. While each of the projects or programmes discussed above possesses many elements of merit, the rural development programmes taken as a totality suffer from a number of shortfalls of which lack of conceptual clarity is the major issue which, in turn, leads to many other problems attending the field of rural development planning in Nepal.

The first problem at the conceptual level has to do with the objectives of a rural development programme itself. While there certainly exists general agreement that the goal is to uplift the standard of living of the people, there is a general lack of understanding as to its operational aspects which include such important areas as its methods, processes and their physical indicators. Most of the time completion of physical projects is taken for achievement. Consequently, despite the heterogeneity of processes and approaches involved in the different projects cited above, an informed and investigative interest in them and a national search for the most desired process are generally wanting.

The second problem at the conceptual level is the lack of understanding of the concept of integration and its dynamics. The fact that problems of rural areas are mutually interdependent and that interventions must be made in several different areas and made in the form of a package is not always widely accepted. This has been demonstrated in the problems faced by the District Administration Plan.

Thirdly, the concepts of people's participation, basic (or felt) needs and sustained growth have not been adequately defined so as to yield useful and concrete policy indicators. To take one example, voluntary labour contributed by the villagers for a project is in most of the cases taken for people's participation, while some hold that it means total popular authority over the ordering of priority and allocation of resources.

Fourthly, the population of the rural areas is considered homogeneous and it is assumed that all will somehow have equal access to the dividends of a rural development programme. In actual fact, however, the interregional and interpersonal differences are immense. But, except in the case of the small farmers' development project, no special provisions are built into projects to overcome the untoward effects of stratification.

As an additional consequence of this social and economic stratification, persons elected to leadership positions in a *Panchayat* or a district are not always beholden to their electors in the exercise of their authority—a fact which has consistently incited the Government to shy away from any substantive programme of devolution of power to local elected institutions. It is also because of this that the country has embarked upon a national back-to-the-village campaign which has the legal authority, among others, to grant or withhold permission to aspiring candidates to contest local and national elections depending upon their continued integrity of character and contribution to local development projects.

Because of these conceptual problems planning for rural development at the local level has taken different forms under different projects. Planning at the local level (in the case of Nepal the district level) would mean the totality of steps which consists of the identification of needs/problems, setting of objectives/targets, identification of potentialities and resources, ordering of priority, allocation of resources, implementation of the programme and its review and evaluation with the purpose of undertaking the cycle of activities on a continuous basis. Looked at from this point of view the District Administration Plan (and also RNRDP because of its DAP leanings) came close to it and the functions of the Government as seen in the Plan were limited to providing policy inputs and resources to backstop the local planning exercise. But such an arrangement would have required several concomitant changes pertaining to decentralization and inter-departmental co-ordination, including turning over the reins of control over their district-level representatives to the co-ordinator at the local level. Some measure of control by the district *Panchayats* could also be expected which would certainly be resisted by the concerned ministries as well as the individuals themselves.

Similarly, other projects discussed above are endowed with elements of local planning processes. For instance, RIDP has a system of local identification of needs and allocation of resources, a system of local review etc. But its weakness in terms of interdepartmental support is too pronounced for any serious interdisciplinary planning effort at the local level.

In the case of the LDD projects, they essentially constitute only part of what could be an integrated whole at the local level. Grants made for allocation at the district level represent a policy of decentralizing resources and authority for substantive planning efforts at the local level. This has already been discussed above in greater detail. But for want of successful integration under DAP, LDD projects in most of the districts remain isolated public works projects.

Under the PDLT programme the local institutions, primarily the village *Panchayats*, are rendered more or less self-reliant in resource endowment for local development. However, while the emphasis of this programme is on decentralizing the revenue powers to the village *Panchayats*, most government functions and technical support resources are located at the district level, thus revealing the implicit lack of harmony in the Government's approach to decentralization of necessary powers and functions.

In conclusion, a local planner in Nepal, whether related to the district *Panchayat*, the CDO or other departmental offices, is confronted with mixed norms. The demands made on the planners vary from project to project and it is a safe proposition for him in any subject to be beholden to the directives of his central department. For reasons stated above, unity of design and approach between departments at the centre still leaves much to be desired.

One additional point needs to be made. It has to do with the low level of motivation of the civil servants working in rural development programmes. Despite measures like additional project allowances and awarding additional credit points towards promotion for working in difficult areas, the quality of dedication to work still needs to be adequately developed.

Ways and Means t Improve the Situation

In order to ameliorate the situation action should be taken at two levels—national and international, the latter gaining in importance because of the increased interest of donor agencies in rural development. Since any international effort is likely to have a demonstration effect on nationally initiated programmes, the donor agencies cannot afford to lose sight of the total national programmes in the countries.

At the National Level

There is no substitute for a thorough national search for the meaning and method of rural development in the country. While this seems to be a truism, the fact that it is not always forthcoming must be acknowledged. Recent writings on successful rural development programmes accord a primordial role to the element of political will. However, it must be realized that no government can afford to overlook the needs of its predominant rural sector and, to that end, a degree of political will should be taken for granted in most situations. In the case of Nepal, the institution of the *Panchayat* system of democracy rising from the very grass roots of Nepali society vouchsafed the primacy given to the rural populace at the political level. Therefore, given the level of political will obtaining in the country, it should be to the disciplines pertaining to rural development to find ways and means to augment a more thorough national search for rural development. The actions and proposals in this regard must take this into consideration and be designed to achieve it.

Institutional development must be taken as the indicator for rural development. The focus of any rural development programme must not be limited to completion of physical targets or projects alone. Policies must be so designed to achieve the emergence of institutions that is a part of the local social structure with the authority to develop and implement plans, generate resources, evaluate and continue this cycle regularly. The fulfilment of this condition will institutionalize the development process at the local level on a self-sustaining basis.

Development of useful technology should constitute one of the foci of a rural development programme. Availability of such technologies provides a strong stimulus for local planning effort and promotes the initiative of the local populace for this purpose.

There should be several levels of local development planning with emphasis on conceiving the level of implementation as the basic level of planning, although no single level should be considered the absolute level of planning. This would entail several different levels of decentralization of government functions with the primary objective of giving the local people the reins of control over the planning implementation of projects that affect them.

The integrated approach must be declared the only approach to rural development. This also implies several levels of integration, the primacy being given to the lowest level of integrated planning, i.e. the village.

Personnel policies should be so developed whereby service givers must be held accountable to service users. In a situation where civil service rules and incentive schemes are inadequate instruments for raising the level of motivation on the one hand, and where the supervision system is either weak or non-existent on the other, such an arrangement is not likely to instil a sense of responsibility to deliver the goods.

Incentive schemes, primarily financial, should be such that would attract people to work in rural development programmes in contrast to other government business. Such an arrangement, in unison with the scheme for local accountability of civil servants described above, is likely to ensure a certain level of motivation and performance among the workers.

Enriching and strengthening local technical manpower should be one of the priority considerations of rural development programmes. Upgrading local technical know-how by continuous training and follow-up should be given priority over deputing technicians from outside the community. In other words, training at different levels must constitute a part of local-level planning for rural development.

A system of review and evaluation should constitute an integral part of rural development planning. Evaluation or review should be done more in the realm of impact than of target fulfilment.

A rational organization for rural development as distinct from subject-matter agencies such as departments of agriculture, education etc. should be established to go into such basic issues as formulating national policies for rural development, conducting training, research and evaluation activities, taking care of interagency co-ordination etc.

At the International Level

Efforts undertaken at the international level have to be nation-specific in that assistance or advice addressed to a specific country should be based on a thorough understanding of the basic problems confronting rural development in that country. So any approach or solution worked out by a donor agency should be thoroughly examined in this light before it is recommended to any given country.

However, there is enough experience to recommend that donor agencies should, as a matter of principle, promote the following issues generally :

(i) Decentralization of government functions to the extent and level that is necessary in order to enable the people to take decisions on projects and programmes that affect them directly ;

(ii) Distributive aspects of development programmes should be specifically emphasized which would require choice of technology, e.g. preference for labour-intensive in place of capital-intensive technology, and implementation of social and economic reform to enable the weaker sections of society to benefit from rural development programmes equally ;

(iii) Institution development should be an integral part of any rural development assistance ;

(iv) Development of appropriate technology should be supported on a priority basis ;

(v) Built-in evaluation (primarily impact evaluation) should constitute an integral part of an assistance programme and should be assiduously adhered to.

It must, however, be reiterated that policies governing assistance programmes should be implemented in such a way as to stimulate and strengthen the process of national search on the issues of rural development in the country.

REFERENCES

1. The Plan provides for maximum and minimum programmes of investment which are predicated on the availability of resources during the course of plan implementation.
2. Immediately following the move restricting the movement of Nepali citizens in India in 1976, the location of small areas underwent a change in favour of the eastern and western frontier district from where people went to India occasionally for employment. Most of these districts are defined as "remote" in the classification of districts in the country.

FIFTEEN

PAKISTAN

Shoaib Sultan Khan

PAKISTAN is predominantly a rural country with nearly three fourths of its population (approximately 75 million) living in about 45,000 villages. The realization that the earlier development programmes had not benefited the most needy in the rural areas made the Government of Pakistan adopt the integrated rural development approach, as a result of the recommendation of a federal committee established to look into the problems of the rural areas and suggest a strategy to assault rural poverty. The committee, which formulated the Integrated Rural Development Programme (IRDP) in February 1972, made an urgent recommendation : "It is of utmost importance to test the programme through pilot projects before replication" The Pakistan Academy for Rural Development at Peshawar, a national research and training institute, was also charged with this pioneering research. The Academy tackled the issue from two angles, one of developing a viable model suitable for replication in the whole country, with necessary modifications taking into account differences in regional conditions, and the other of documenting, evaluating and readjusting the strategy as a guide to researchers and planners alike.

Components of Integrated Rural Development Programmes

In the light of the field experience of the Academy in the Daudzai Project in North-west Frontier Province, the basic components of integrated rural development programmes at the local level can be listed as follows :

(a) To establish a centre (*Markaz*) of development administration for a conveniently small area (a police station area including about 70 villages with 70,000 to 100,000 population). To construct there an office building and a residential colony to accommodate all departmental representatives. To promote co-ordination between the departments and local institutions such as co-operatives and self-government;

(b) To establish a local government council at the *Markaz* level and to locate the *Markaz* council at the centre and co-ordinate its work with the departments through a Local Government Department official (at present called project manager) and an elected chairman of the council;

(c) To plan and complete the physical infrastructure for the area : roads, drainage, irrigation, land improvement and electrification. To appoint for the purpose an engineering unit for survey and blueprints, to prepare the plans in consultation with the village people and to train and supervise local organizations (co-operatives and councils) for executing the works and maintaining them permanently;

(d) To make organization and mobilization of the village people the chief concern of the centre. To create for this purpose a two-tier organization—village organizations federated at the centre;

(e) To train intensively and continuously many cadres of local leaders (organization managers, model farmers, project committees, councillors, teachers, paramedics etc.);

(f) To promote extension education and servicing by all nation-building departments through village institutions, especially the two-tier organization. To train continuously for this purpose representatives of organized village groups as extension agents (managers, model farmers, paramedics, youth club leaders, schoolteachers etc.);

(g) To create a strong and viable structure of co-operative credit for the majority of the farmers (most of them small) in the shape of two-tier organization—primary societies federated at the *Markaz*. To promote thrift, saving and investment along with credit, to collect equity capital and aim at self-financing and self-management. The *Markaz* federation's functions will be to co-ordinate, train and supervise, along with banking.

The village organization, in most cases, will be entrusted with the execution of infrastructure work;

(h) To add gradually other components (e.g. machine workshop and tractor station, tubewell drilling and maintenance section, co-operative processing unit, agro-industries). The emphasis will be on the growth and expansion of the co-operative sector financed and controlled by the majority of small farmers;

(i) As the villages become better organized and more productive, to expand the health, family planning and education programmes, e.g. establish a school board to organize—along with the existing institutions—literacy, youth and women's programmes;

(j) To integrate the local-level plan with regional and national plans.

Economic, Political and Social Objectives

These programme components in a specific manner are related to broader economic, political and social objectives which are being emphasized by the Government. For instance, the economic impact of the *Markaz* programmes will reduce losses caused by the absence of protective infrastructure and will increase productivity by at least 30 per cent in five years. They will also provide more employment on a temporary as well as on a permanent basis for at least 20 per cent of the labour force.

They will raise agricultural skill and yields by 25 per cent in five years. They will build co-operative equity capital belonging to small farmers, teaching them fundamental habits of thrift, saving and investment. Politically, local government institutions will promote the exercise of popular control and responsible management and leadership. The two-tier organization will strengthen the small farmers' position and reduce the excessive privileges acquired by large owners of land and capital. Socially the work on and the development of infrastructure will improve the condition of wage workers. Thus tensions may be reduced through wider employment, increased productivity of land and the stronger position of the small farmers.

It is obvious that a programme of rural development is a

long-term activity. Everything cannot be achieved at once. Organization takes time, so does the construction of the physical infrastructure. If the initial work is done thoroughly, though slowly, there will soon be a gathering momentum. The capacity for more and better work will grow rapidly. This lesson was well illustrated in the Daudzai Project.

Fundamentals for Development

The Academy research showed that as far as small farmers are concerned, while many factors are involved, there are three fundamentals for development :

(a) It is not possible to improve land, to extend irrigation, or control erosion or waterlogging except through the joint action of disciplined groups;

(b) It is not possible to develop an effective extension system for small farmers except through the same disciplined groups;

(c) It is not possible to develop an effective credit system for small farmers unless they are organized in disciplined groups.

Accumulation and investment of capital, acquisition of more productive skills and protection and improvement of land are the fundamental factors on which depends the economic advancement of the small farmers and by which their progress can be measured. Other factors, such as the presence of experts in the form of government extension services, or facilities for health and education are indeed helpful but their effectiveness increases or decreases in proportion to the discipline or anarchy prevalent in the villages.

Phasing and Funding

Since the Daudzai Project was developed with the sole objective of expansion and replication of the model to the rest of the North-west Frontier Province in particular and to the rest of the country in general, a programme based on the Daudzai experience was prepared and put into execution early in 1975. In the light of the field experience, the future phasing and funding of the rural development programme in the North-west Frontier Province was planned as follows :

First year

1. Posting of the following staff :
   (i) Project manager and assistants
   (ii) Co-operative organizer assistants
   (iii) Engineer and assistants
2. Completion of *Markaz* building and residential colony
3. Engineering survey and preparation of plans and blueprints for infrastructure works
4. Co-operative education and organization for works, agricultural development and co-operative credit

Second year

1. Posting of department officers
2. Registration of village organization and federation
3. Infrastructure works, 10 per cent
4. Agriculture extension and co-operative credit for 25 per cent of villages
5. Establishment of local councils

Third year, fourth year and fifth year

Increase of 20 to 25 per cent in each activity, viz., works, extension, credit etc., to cover the entire area in five years. The funding of each *Markaz* for five years was projected on the following scale :

	(thousand of rupees)
1. Buildings	700
2. Equipment and transport	150
3. *Markaz* staff (local government)	736
4. Department staff	836
5. Grant to federation	400
6. Infrastructure works	3,200
7. Co-operative credit	800
Total	6,800

This was the general level of investment prescribed for every *Markaz* area.

Expansion and Replication

By the end of 1975 the Daudzai approach had been extended

to 34 *Markaz* areas (including 5 *Markaz* areas) taken up under assistance from the Government of the Federal Republic of Germany, covering nearly one third of the province. The recruitment of the core staff of these *Markaz* was done by an *ad hoc* committee headed by the chairman of the Provincial Public Service Commission, pending regular recruitment, and a four-month training course devised by PARD was given to the staff before their posting to the *Markaz* areas. A monthly conference of all the *Markaz* areas core staff was also held at the Academy, after the initial training, to monitor their work and obtain the necessary feedback. A directorate of integrated rural development was set up at the provincial level to execute the provincial programme.

Diagnostic Survey

In Daudzai the entire project staff was involved in preparing surveys and blueprints—first without consulting the local population and then with full consultations. The final surveys and blueprints revealed that there existed significant differences between whatever limited data were available (mostly from revenue records) and what was finally surveyed. Also, the blueprints made after consulting the local people were found to be more realistic and beneficial than those planned independently by the project staff.

Accordingly, while replicating the model in other *Markaz* areas, the diagnostic survey began in the first year and lasted through a full year. It provided a detailed picture of villages in a *Markaz* area at a micro level, generally not available in government records. The survey included basic statistics of the area (area, population, crops, cropping intensities, yields, land holdings, schools, dispensaries etc.) and the likely infrastructure works schemes which would benefit the rural population living within the *Markaz* area. The blueprints for works included : the cost-estimates; the likely benefits; and the number of beneficiaries. The survey and the blueprints in a *Markaz* area became : (a) the basis of the works programme on which PRs 3.2 million are to be spent from the second year through the fifth year of the programme, or 47 per cent of the *Markaz* budget for five years; (b) the mechanism which focuses on the real clientele (the beneficiaries); and (c) the basic plan through which the various *Markaz*-posted departmental functionaries design their respective departmental strategies. The survey and blueprints were prepared in consultation with the people. The benefits of preparing surveys and blueprints for each *Markaz* are evident in that the rural development strategy for each *Markaz* area will be unique depending upon the local problems and priorities. Planning for such problems and priorities will not be possible without the area blueprints and surveys.

In this area the constraints included the development of a corps of rural engineers; misuse of funds and a reluctance towards evolving plans in consultation with the local population. To surmount these difficulties the methodology of having a pool of engineers at the district level for supervision of the work of sub-engineers at the *Markaz* level was evolved. In cases of corruption, instead of withdrawing powers and centralizing them, the remedy adopted was to bring to book the corrupt functionaries promptly; and to counter the tendency towards imposition of plans on the people, the monthly meeting of the core functionaries of the *Markaz* was initiated.

Works Programme

As noted earlier, works schemes are based on the surveys and blueprints prepared during the first year of a five-year programme. The major portion of a *Markaz* budget is expended on this activity, resulting in direct visible benefits from land development projects and the building of physical infrastructure. It is important to note that works schemes are not undertaken till blueprints have been prepared and village organizations formed. In fact, no works scheme could be implemented unless a resolution had been passed by a village organization and project committee nominated by the organization to undertake the scheme in consultation with the project engineer. This implies that only those works could be implemented which would bring collective benefits. The first priority in works schemes was given to projects leading directly to increased agricultural output or leading to income generation. The prevalent system of getting projects executed through contractors was substituted by getting it done through village organizations or their nominated project committee.

Unlike Daudzai, funds for works programmes in other

areas were not made available as stipulated under the plan and the directorate at the provincial level tended to centralize the project-sanctioning authority. A system of pre-audit was also imposed on the works programme funds, which caused delay in release of funds for individual projects resulting in consequent postponement of the execution of the project. An attempt was made by the Academy to resolve these issues through the intervention of policy-makers, planners and administrative heads.

Village Organizations

During the first year, the local population in the *Markaz* area is exposed to the rural development staff (*Markaz* core staff) conducting surveys and preparing blueprints. The first year is a period of rising expectations and development awareness. Despite the temptations of undertaking investment in works schemes, no physical infrastructure gets built. In the second year, a two-pronged rural strategy begins. Pressures built earlier by rising expectations and development awareness are harnessed through the formation of organizations (co-operatives). In fact, village organizations become the pivot of the entire rural development programme. Co-operatives of village organizations as treated in the rural development strategy in the North-west Frontier Province are "local development organizations" at the village level and do not have any similarity with the traditional co-operatives functioning in the subcontinent for the last 50 years or so. Establishment of these organizations, that take over the development role, the maintenance of works schemes and become the vehicle for departmental extension programmes, is the base on which the whole edifice of rural development is built up. Once the village co-operatives are formed and registered under the local law, these are then federated at the *Markaz* level. The federation at the *Markaz* is at the apex of the co-operatives and also functions as the central bank for co-operative capital.

The response of the rural population to this offer of partnership between the Government and their organizations was positive in all the *Markaz* areas. If there was any slackening of effort in this regard it was more often on the part of the

Government than the people. It also resulted in the establishment of a development-oriented local leadership.

Credits and Loans Programme

It was noted earlier that the real clientele for effective rural development is the majority of farmers who own small land holdings. The commercial/agricultural bank credits for agriculture development have, regardless of the intention, concentrated primarily on farmers with larger land holdings. The small farmer is seldom exposed to credit for agricultural inputs. Since the present programme focuses on small farmers who are less exposed to modern agricultural inputs, it is evident that this service could only be provided through group action. Loans are, therefore, given to members of village co-operatives for seeds, fertilizers and pesticides. Collective loans are also given to co-operatives for pesticides, spray-pumps, soil-testing kits, tube-wells, etc. An interest rate of 12 per cent per annum is realized. These interest rates make the small farmers realize that such credit is not charity. An amount of PRs 800,000 will be available to each *Markaz* during the five-year phasing period as capital for credit and loans or 11.7 per cent of the total *Markaz* budget for five years. This amount will be repaid to the Government in eight instalments after the sixth year of the *Markaz* functioning. Therefore, the credit and loan amount for each *Markaz* is an interest free loan from the Government to the *Markaz*. The equity capital and savings accumulated in the weekly village co-operatives' meetings during the five-year *Markaz* expansion programme is anticipated to make co-operatives and the *Markaz* federation independent of government grants and loans.

Despite reluctance on the part of the Provincial Co-operatives Department to register the *Markaz* federation as a central bank, and inadequate and irregular provision of funds, the credits programme in most of the *Markaz* areas made significant inroads and impact. The rate of repayment was highly satisfactory compared to that of the loans advanced by credit institutions.

Departmental/Extension Programme

The programme aims at providing the services of nine departments at the *Markaz* level. The functionaries physically posted at the *Markaz* complex come from the following Departments : Revenue, Health, Education, Animal Husbandry, Forest, Agriculture Co-operatives, Small Industries and Electricity. The primary objective is to train, through departmental functionaries, representatives of village co-operatives in their respective fields. In addition, the functionaries carry out their respective departmental development programmes. This mechanism on the one hand benefits not individuals but groups, while, on the other, puts less pressure on the available thin layer of a government staff. The extension benefits are therefore maximized in relation to inputs. The strategy envisages a *status quo* regarding the development efforts of the various government departments. All nation-building departments continue to pursue their respective development plans in the rural areas. However, through the *Markaz* their efforts will be more effectively tapped by a thrust in which rural population (through co-operative action) will take the initiative of "demanding" a project or service and the departmental functionaries will be "supplying" the project or service. The project manager or his staff will be the co-ordinator between the "supplier" and the "consumer". This will :

(i) Eliminate fears that the *Markaz* is a super department ;

(ii) Keep the departmental functionaries' seniority and promotion intact with their respective departments ;

(iii) Eliminate interdepartmental jealousies so frequently cited as one of the causes of failure of earlier programmes.

Despite every effort to allay apprehensions of the line departments, it needed an Herculean effort really to convince them of the intentions of the programme. There was some change perceptible but the goal appeared still a long way off.

Conclusion

Inevitably, a programme of the type that innovates, deviates from standard procedures and aims at a particular group in the economic system will encounter opposition. It is a long-time effort. It aims at a fundamental behavioural change. Although results in economic terms can be shown in the short run, the ultimate test of success is whether the programme can change the way the village people think about themselves and their environment. Unlike large engineering projects where scientific principles are well established and international experience has accumulated, there is no comparable body of knowledge for rural development, and the programme based on the Daudzai approach is in this sense experimental. There will inevitably be unsuccessful experiments as well as successful ones. This process of experimentation needs time and cannot be hurried excessively. For this reason it must have support for a sufficiently long period to ensure full opportunity to succeed. Political leadership cannot insist on definite performance in the very short run so that it will have an impact within a particular term of office or before a particular election.

SIXTEEN

THE PHILIPPINES*

Benito F. Estacio Jr.

Past Experience : A Summary*

UNDER the broad umbrella of the National Economic and Development Authority (NEDA) several approaches have been initiated in the Philippines for planning rural development on a decentralized basis.

For instance, in order to promote more balanced development among the various regions of the countries, 13 regional councils were established in 1972 to formulate regional development plans and supervise their implementation.

In 1973, the Cabinet Co-ordinating Committee for Integrated Rural Development Projects (CCC-IRDP) was established to select priority areas for integrated development, prepare project studies, negotiate funding and supervise implementation. The main criteria for selecting those areas were their relative status with regard to underemployment and their potential for rapid development. Integrated projects of this type were launched in Bicol, Mindoro, Cagayan Valley and Samar.

As regards the non-priority areas, they were left to the initiative of the sectoral planners. For instance, the Department of Agriculture has been promoting "packages" of practices for improving rice, corn and livestock production. The Department of Agrarian Reform has co-operated with other line departments to secure the required services for new owner-cultivators; it has provided such services directly in resettlement areas. For small irrigation projects, the policy is to

* Summary prepared by the ESCAP secretariat.

promote small farmers' co-operatives called *Barangay* Irrigation Service Associations (BISA); under this programme, pump or gravity feed irrigation systems are developed together with improved production packages, marketing and processing services are provided, as well as assistance for mechanization.

Currently, the Department of Local Government and Community Development has encouraged the creation of provincial development staff to develop comprehensive provincial plans. This Department is also responsible for supporting the local government structure (Municipal Councils, *Barangay* Councils) and promoting the development of a pre-co-operative system (*Samahang Nayon*) for local farmers.

Current Practice

The rest of this paper is devoted to a detailed description and analysis of current approval to integrated rural development in the following three examples of local-level planning.

The Barrio Irrigators' Service Association

Background

The *Barrio* Irrigators' Service Association (BISA), was organized in 1973. The programme aims at increasing the agricultural production and income of the farmers in the project areas by channelling the technical and financial resources available from the various government and private agencies and institutions involved in agricultural production programmes through the Irrigators' Service Association (ISA).

Under this plan, the farmers within the technically-feasible project areas, that is, ricelands in need of irrigation with a nearby dependable surface water supply and preferably within the service reach of an electric co-operative, would be organized into ISAs.

These associations would, in turn, set up their own pump irrigation system. The farmer-members would then undergo effective production and management training programmes which would enable them to optimize fully the utilization of their irrigation facilities as well as individual and organizational resources in their production activities. These ISAs would serve as the unit cells from which the Government might harness the production potentials of the farmers by way of utilizing the association as a channel for technical and financial assistance.

Objectives

Specifically therefore, the programme is intended to achieve the following objectives for each project area :

(i) Organization and development of a farmer's co-operative, called the ISA, which will build, operate and maintain and manage its own irrigation system which would provide irrigation services to farmer members; and plan and implement agricultural production projects which would provide other essential services to its farmer-members;

(ii) Construction of an irrigation system using either electrically-driven pumps (in project areas covered by electric co-ops) or mechanically-driven pumps (in other areas). The setting up of the irrigation system, construction, ownership, operation and maintenance and management are to be carried out by the farmer-members of the association;

(iii) Introduction and implementation of improved farming and marketing practices.

Programme Concepts

In order to attain the aforementioned objectives, institutional activities in the organization phase come in terms of information dissemination, briefings, organization of the Provincial Irrigators' Teams (PIT) the Municipal Irrigators' Teams (MIT) and through training, conferences and workshops, recruitment and membership campaigns, organizational meetings as well as communications and information support activities. Its technical organization includes undertaking of area surveys, preparation of designs, estimates and feasibility studies, construction of power lines and physical irrigation system structures, water rights application as well as financing activities. The latter activity is a concerted work of the National Project Team, the National Irrigation Administration, the National Electrification Administration, the Provincial Irrigation Team, the Rural Electric Co-operatives and the Irrigators' Service Association. Upon organization and construction of these various *barrio* associations and irrigation systems in selected areas of the country, training programmes in management skills, production skills, and project development are established.

The management skills training is aimed to strengthen the ISA organization to undergo a series of training activities geared towards enabling them to acquire the concern and skills for managerial functions of the association, that is, from planning, organization, control and evaluation. The participants in this training programme are the Board of Directors, the chairman and members of the Irrigation and Management Committee, the Inventory and Audit Committee, the Grievance Committee, and four members coming from the ISA group.

The production skills training is geared to enable ISA members to acquire or adopt improved varieties of rice seeds, farm and marketing practices. The training is done through a series of lectures and field work demonstrations related to pre- and post-harvest activities.

The project development training is a series of planning workshop and field work training activities. This is geared toward enabling the trainees to prepare both education and economic plans and for the associations.

The implementation of the BISA development programme was initially introduced in 16 provinces across the country. By 1975, there were 318 irrigators' associations with an estimated membership of 15,811 farmers. The total average services by the programme as of 1975 was 62,250. The estimated total harvest for the fiscal year 1975-1976 was placed at 2,213,750 cavans of rice. This programme later evolved into a corporate system presently known as the Farm System Development Corporation.

Discussion

The success of the BISA programme is due to the following factors. The initial stage of project conceptualization was carried out by farmer-technicians employed by the Development Academy of the Philippines, a semi-government agency. These young men and women had many years of experience in various forms of co-operative endeavour in their mother agencies before joining the Academy. Their knowledge of the fields, their sensitivity to the peoples' needs and aspirations and their technical know-how were fused into one and directed toward an integrated planning approach. They all understood and co-operated in the project formulation and later as participant observers of the newly established organization of the Academy.

While the planning exercise was done in Manila, this was, however, declared as an open plan for consideration subject to validation in the field by conducting field visits and consultations with the target beneficiaries. Acceptability of the project plan was facilitated through informal lectures at the *barrio* level. Having prepared the people for this novel approach at the organization of an Irrigators' Service Association, plans were presented, revised and corrected by the farmer-beneficiaries.

Reformulation of project plan was made with reference to peculiar substantial change in concept as well as peculiar approaches in carrying out a plan. Short cuts were made and at times more demanding on the part of the project managers as well as townspeople. Research and communication staff was competent in handling both information dissemination and monitoring of activities on a monthly basis.

While the programme may be viewed as a success, the following were some of the essential problems identified in the field :

(i) Since project conceptualization was initiated in the central office in Manila, its level of sophistication in terms of targets and applicability were painstakingly revised by the target beneficiaries. Functional details of the plan became a reality only at the *barrio* level;

(ii) Institutional support in most regions as well as at the provincial levels were not made readily available to the Provincial Irrigators' Team and the field workers at the municipal and *barrio* levels. A series of information campaigns, training and other approaches were found taxing on the part of field workers in priming the communities in the prioritized region and province;

(iii) While the information dissemination and training were conducted among the projected participants and *barrio* Irrigators' Association, a majority of the farmers took a wait-and-see attitude. Planning, project implementation and actual construction of the *barrio* irrigation system had been found very difficult to accomplish at the initial phase of the project. This could now be attributed to the following factors :

(a) Psychologically, the farmers in the *barrio* had not been altogether sold to the new approach of a *samahan patubigan* (irrigators' association), specifically towards the construction, management, maintenance and control of a mechanized irrigation system. The individualized gravity irrigation (where water was abundant) was made use of by traditional wet-rice farmers;

(b) In some areas, climatic conditions and physical environment were not conducive to an irrigation system that was pronounced by local leaders as "doomed to fail";

(c) Farmers' expectation of change agents had been limited to past experience where government agencies delivered the finished product of "development" without citizens' participation;

(d) Delivery of irrigation pump components, motors, engines, accessories and piping was subject to availability of posts, accessibility of the pilot area and the unpredictable weather conditions in most areas of the country. This was due to lack of suppliers who could deliver the requirements within the scheduled period of the project which, in turn, was because of the lack of qualified technical manpower in the provinces covered;

(e) There was delay in technical surveys, designs, and estimates, completion of which was delayed by more than two months;

(f) Lastly, the failure of the provincial government to put up the expected provincial counterpart support. This hampered the operational activities of PIT and MIT.

The BISA programme may be dubbed one of the few success stories in the Philippine local-level planning experience. The method used for formulating and monitoring revised local plans, results achieved and difficulties encountered in the fields was plain logbook keeping. Daily experience in the field was entered in diary form and a monthly resume sent to the central office for evaluation. A random sample of 11 provinces was surveyed and spot checked and intermittent ocular visits were carried out by the National Irrigators' Team. The results of survey findings, spot checking and field visits, and the monthly report of the action men from the field became the bases for reprogramming and adjustment of expectations, deadlines and for repackaging of proposals to the National Irrigation Administration.

The National Irrigators' Team, composed mostly of graduates from the State University College of Agriculture, were very effective managers and implementers of the programme. Local development communication experts were utilized and added the much needed and essential component of the programme. Linkages with the regional, provincial, municipal and *barangay* councils were tapped through an excellent communications network.

In the context of the aforementioned premises, without the regular red tape of the Philippine government agencies, the BISA programme would indeed qualify as a success story. This case is presented here, not so much to indicate how to succeed in local-level development planning as to explain the nature and seemingly simple structure of local-level planning. Before others may think that this success story is that easy to accomplish, there is a need to explain the political atmosphere and important decision-making that enhanced the success of the programme.

Firstly, the National Irrigation Administration had been deeply concerned about the ways of maximizing its services to the rural areas.

The much needed component of the programme-monitoring and evaluation system, linkages with the regional, provincial, municipal and *barangay* councils, was established through a network of Irrigators' Teams.

The crux of the matter lies in the *barangay* or community-level participation where local resources, both human capabilities in water management and indigenous materials, were available. Each irrigator's team entertained the field technicians, listened to their plans and adapted those plans of action that were complementary to their existing indigenous family-level water management and water rights custom-law. In some areas there was total reaction to and co-operation with the BISA programme on account of an already existing condition where the entire *barrio* had been waiting to accomplish (but could not itself organize) a system of linkages, and the absence of large capital investments. In both instances, a systematic organizational network becomes a viable enterprise where farmers at the community level join together and formtheir own policy-making body, related itself to the over-all system of relationship to the provincial, regional and national organizations for exogenous assistance in both complementary technical and financial aspects. The economic benefit derived from this co-operative endeavour at the *barrio* level became operative. Where the *barrio* dweller in the past took care of his own plan for irrigation (family level), the benefits to be derived if he were to join the pool in turns indigenous know-how and resources complemented by an exogenous assistance now became clear to him. There are, however, existing co-operative endeavours at the *barrio* level such as those reported in Ozamiz, Mindanao and Lewis of the Mambubuyong wet-rice farmers of the northern Philippines. These existing indigenous organizations were reviewed by BISA research teams and lessons from their successes and failures incorporated in the NIA planning effort.

The Pantabangan Resettlement Project

Background

The Pantabangan Resettlement Project was an offshoot of the construction of the largest dam in Asia. The Pantabangan dam is a major feature of the Upper Pampanga River Basin Project designed to provide a reservoir with a storage capacity ot 2,996 million cubic metres. This project, started officially in June 1969, is a massive engineering undertaking for the purpose of providing year-round irrigation for rice and other diversified crops, domestic and industrial water supply, as well as the much needed hydroelectric power. It was projected that the economic benefits to be generated by the project would be millions of pesos annually. The direct and proximate beneficiaries of the project are two cities (Cabanatuan and San Jose) and some 14 municipalities.

The construction of this dam, however, necessitated the submerging of the town of Pantabangan and several other *barrios*. Some 4,000 hectares of prime residential land would have to be submerged and some 1,600 households displaced in areas designated as relocation sites. These areas are non-contiguous.

According to a study by the Institute of Environmental Planning of the University of the Philippines, the process of relocation of households affected by the project was to be carried out with all precautionary measures in order to cushion the impact of the transfer. It was projected that the actual relocation of families in several phases was to take place between March 1973 and March 1974.

In preparation for the relocation, a series of community meetings and consultations were held to explain the merits of the proposed government projects and the promise of a resettlement programme. The resettlement programme included six major components.

The first was a compensatory scheme where loss of properties was paid and loss of work opportunities would be provided for the corresponding manpower skill training to help the settlers cope with a new life style. Payments on lost properties were based on the schedule of appraised values established by the Provincial Appraisal Committee.

The second was the distribution of farm and home lots to each household head of the settler-relocatees. Following the Department of Agrarian Reform schedule, a total of 2.5 hectares of irrigated land was to be awarded to each family or a total of 4 hectares for non-irrigated areas. The minimum area for a home lot was 600 to 1,000 sq m. All land distributed to relocated settlers was leased by the Philippines Government for a period of 40 years, subject to renewal for the same period of time. However, before actual transfer, most of the available home lots had already been located.

The third component of the programme was direct assistance from the National Irrigation Agency (NIA) for housing. The first option was a choice between three model core houses constructed under the project payable for within a period of 13 years in equal monthly instalments; the second was reconstruction of some usable materials from their old houses with new construction materials available on loan also payable in 13 years; the third was total reconstruction of houses using the old materials at no cost.

The fourth component was sustained food assistance for a

period of three years, the rationale behind which was the length of time before farmers could cultivate, plant and harvest in reasonable quantity sufficient for their year-round subsistence. Food assistance was made available through the United Nations World Food Programme. Rations included wheat flour, rolled oats, corn-soya milk and corn and sorghum for livestock feeding.

Financial assistance was the fifth component, providing some soft capital loans for the relocated farmers. By virtue of Presidental Decree No. 35, the sum of 20 million pesos was released for this purpose. Under this scheme there were three types of loans made available, namely, housing, subsistence and project loans. All loans were administered by NIA and the maximum amount available was £ 6,000 while the minimum was £1,000.

The final component of this programme was the pool of resources from various government agencies. Prominent among them were NIA and DAR. Twenty other government agencies joined in as their services were tapped by the interagency committee.

While the above programme components were viewed by the relocated settlers with eagerness and high expectations, implementation and operationalization were sadly lacking. There were unforeseen variables — social, economic, physical and administrative — that developed into nagging issues and problems for the implementing agencies.

In May 1974, the Development Academy of the Philippines was requested to look into the situation and define its area of participation. As an aftermath of a series of field visits, interviews and a cross survey of three separate relocation sites, the Academy convoked two workshops-seminars participated in by both settlers and representatives from various co operating agencies. The main objective of both workshops was to identify the existing problems of their respective relocation sites and to propose some solutions. Among the many problems identified and given priority by the participants were the following:

(i) There was non-congruence in terms of work expectation between the old and new sites. Farmers were unprepared to respond to the demands of a new environment;

(ii) The relocation sites, being rolling, were insufficient

agricultural land dubbed to be cultivable with maximum benefit of their labour and other farm inputs;

(iii) Ironically, where the relocation sites were surrounded by the massive dam and its steady supply of water, there was low supply of domestic water to the settlements;

(iv) The absence of terracing brought about heavy erosion to farms as well as home lots during heavy rains;

(v) The zoning pattern was not strictly observed by some relocated settlers.

In response to these problems, the Development Academy of the Philippines was asked to come out with a development framework plan for the entire resettlement project. The plan was to provide strategies of action that encompassed the series of interrelated projects for the promotion and management of the socio-economic and physical growth of the resettlement areas.

The specific objectives of the task were the following:

(i) To resolve the interagency conflicts over management of different development aspects of Pantabangan;

(ii) To recommend alternative management systems and plans;

(iii) To assist DAR, NIA and other co-operating agencies in relating and integrating land development, infrastructure and social services projects;

(iv) To identify possible economic activities that would help settlers attain self-sufficiency enabling them to adjust economically and socially to their new environment.

The planning approach adopted by the DAP team was to deviate from the traditional phase of planning concerns. The rationale behind this was the fact that there was an urgent need for action projects and that the settlers had been test-wise because of over-exposure to survey teams in the preceding years. The use of actual visit and panel interviews of key informants (both settlers and agency representatives) was resorted to. Baseline data were gathered from the various government agencies who made their own surveys. Analyses of secondary and primary data were guided by the conceptual framework to which the task was addressed.

As a result of this supplementary task, specific planning

strategies adopted for each of the three major areas of developments concern were formulated.

The Socio-Economic Plan

The strategic framework plan redirected the relocation programme to a process of developing a set of mass-based, employment-generating projects to serve as the core activities within each village-community in the new relocation site. Each project was to be designed as a complete package, defining production and technical requirements, possible marketing tie-ups, a financing scheme, organizational and management structure, as well as other requirements for immediate and effective implementation.

The rationale behind this move was the fact that the settlers could not definitely rely forever on the hand-outs programmed for them from the very beginning of the relocation.

The proposed agri-economic projects were classified as agricultural and medium- and small-scale industries. A set of agricultural projects was identified by the settlers of each settlement, packaged and conceptualized in the language of the implementing agencies by the DAP team for their consideration. Among the agricultural projects presented to NIA and DAR were coconut with intercrops of upland rice, indigenous (sweet) potato and manioc, abaca, banana, bamboo, vegetables, sunflowers and herbal medicine gardens. The crops are indigenous to the area (except for sunflowers) and were thought of as a carry-over of the traditional practices and interests of the settlers. These packages were given priority by the settlers to serve as their basic means of subsistence and projected to be long-lasting, income-generating activities.

In the field of small-scale industries, the three settlement areas preferred to develop bamboo crafts and garments manufacture. *Ipil-ipil* would provide a steady supply of leaves for pellets for animal feed, fertilizer, firewood, and seeds, and would provide the natural reafforestation much needed in their new found place—a virtual green desert covered with *cogon.* One of the more interesting projects identified and packaged was the processing of rabbit pelts. Rabbit raising had been recently introduced in the old site of Pantabangan but people thought if they could avail of backyard raising it might be widely propagated in the five settlements.

Rabbit raising was expected to provide a steady supply of meat and realize an attractive income from the skins when properly processed and marketed, both domestically and for export purposes. There were other income-generating projects, like mushroom culture (button mushroom or *taingang-daga*), that were specifically packaged for the five settlement areas of Tanauan, Palayan, Lublub, Marikit, and Decalay.

The Physical and Land Development Plan

Meanwhile, the physical development of the resettlement sites was evaluative in nature. This is, each resettlement site was pre-planned by the implementing agencies but there was a need to review such plans based on questions related to soil erosion, land distribution, land use, zoning and the like that were raised by the settlers during the tripartite consultation. The proposed physical development plan was therefore focused on the study, evaluation and correction (where necessary) of existing physical developments and unimplemented land development projects. This was necessary pending the resolution of certain problems and issues regarding distribution and management of agricultural lands in the relocation sites.

The physical plans prepared were intended to integrate and put in order land development, infrastructure constructions and provision of social services sites. The main concern of the DAP team was to provide in the design the broad concept of the settlers' identity with their new community through an orderly and well perceived strategy for growth and through careful location of community facilities and amenities. Throughout these planning efforts, community leaders were identified and requested to join the planning board for this physical planning exercise.

The Settlement Management Plan

In an attempt to provide an efficient vehicle for the management of the resettlement areas, the functions and activities of all government agencies as well as private institutions involved in the Pantabangan programme were reviewed and analysed. The main criteria for this review were organizational structure; type of services rendered; degree of participation and budget allocation capability and manpower involvement.

Based on this study, the proposed management plan of the settlements provided the strategy of approaching management in two administrative tiers or levels of services. That is, the first level consisted of the administration of the resettlement programme by DAR, targeted for November 1975. The second consisted of the integration and co-ordination of the functions and activities of various government agencies.

The first level of administrative services, consisting of DAR resettlement activities, required a clear delineation of the functions and activities turned over to DAR and NIA. Although a supplemental memorandum defined the general functions of each agency, certain guidelines had to be established for each agency, to promote efficient turn-over of functions and to prevent future problems between the two agencies concerned.

The second level of administration involved the creation of an interagency co-ordinating committee. As constituted, it is intended to co-ordinate the various activities in the resettlement areas as well as the participation of the specific agencies. It is not envisaged that it will pre-empt any or all of the work activities of the settlers, neither does the committee intend to be a permanent entity. It shall, however, seek to introduce and initiate opportunities and guide settlers in establishing a self-supporting community in each of the resettlement areas.

Discussion

The initial Pantabangan experience is a classic example of planning from above with the good intention of relating to the would-be beneficiaries. However, good intentions are not enough. Development planning, control and management as suggested in the over-all Pantabangan resettlement framework plan and carried out by the DAP staff spelt out some strategies on project management, physical and socio-economic development. The crux of the matter was in its definition in detail of how the plan was to be operationalized. This was beyond the task of the planning team. The implementing agencies, on the other hand, failed to see the need to translate such strategic plans into more concrete project components. This failure may perhaps be attributed to human factors in the interlocking agencies rather than to lack of technical competence. The definition of roles and responsibilities are not clear cut. There are inconsistencies with regard to the level of the roles and functions of the National Irrigation Administration and the then Department of Agrarian Reform. As a result of these sociological conflicts between co-operating agencies, the ultimate victim of all sophisticated planning endeavors is the target beneficiary.

The millions that had been poured into the project went to various government as well as private agencies.

The degree of people's participation in the exercise was not very extensive. Consultations were held but technological innovations, as in the introduction of giant *ipil-ipil* and other medium- and small-scale industries such as pelt-making, were exogenous in character. Community leaders are quick in the identification of such project components but only a few really understand the rationale behind all projects where there seems to be indirect bearing on their pressing problem—that of survival in a hostile environment.

The physical plan was indeed a product of bilateral consultations between the DAP planners and the community leaders specifically for the design of each settlement site. The futuristic approach of planners, however, led to some confusion and untimely raising of people's aspirations in terms of goals and targets which could not be accomplished overnight. Specifically, the proposed physical development plan spoke of urban areas consisting of residential, industrial, commercial, institutional and recreational areas. As a result of this consultation process, speculation followed and, since there was limited agricultural land in each of the relocation sites, this added confusion to issues previously unresolved.

Meanwhile, the interagency committee could not cope with the interpretation of plans and the individual problems of each settler. There were just too many agencies involved and co-ordination became unwieldy.

The Proposed Philippine Virginia Tobacco Administration Integrated Programme

Background

The Philippine Virginia Tobacco Administration, created in

1960, primarily mandated to promote the effective merchandizing of Virginia tobacco and to establish and maintain a well balanced production and consumption of Virginia tobacco and its manufactured products. It is further mandated to improve the living and economic conditions of the people engaged in the tobacco industry, specifically, in the provinces of Abra, Ilocos Norte, Ilocos Sur, and La Union.

For the past 18 years, PVTA has developed and implemented some programmes and projects focussed on the marketing and trading of Virginia tobacco. Tobacco technicians and agricultural extension workers were hired and fielded to the four provinces to meet the demands of the farmers in tobacco-growing and marketing in particular.

Lacking a systematized programme plan that will integrate the interlocking relationship between agricultural and technological services that will cater to the three main activities —from tobacco production and processing to trading and marketing—PVTA has reviewed its existing programmes and projects vis-a-vis internal and external capabilities and resources. This review has brought to the fore some of the demands and problems requiring attention. Among these are the following :

(i) The need to reorient the internal processes of PVTA to the new directions it wishes to take;

(ii) The need for new skills in planning, project identification, design and management;

(iii) The need for appropriate technology to implement various agricultural packages apart from Virginia tobacco but essentially related to its sustained industry;

(iv) The need for better qualified extension workers to carry out projects;

(v) The need for continuous and systematic monitoring and evaluation of client needs and of projects implementation.

Based on the foregoing needs, PVTA has entered into a joint planning task with the integrated Area Planning Department of the Development Academy of the Philippines to develop a programme plan that will embody a system to ensure the continuity of PVTA projects, specific and responsive to the local needs of each of the four tobacco-growing provinces.

Goals and Objectives

Thus the proposed Five-Year Integrated Virginia Tobacco Agricultural Development Programme will spell out project components that will :

(i) Strengthen PVTA internal capabilities in directing the growth of the Virginia tobacco industry in the Philippines;

(ii) Identify appropriate programmes and long-term goals and objectives;

(iii) Design specific action plans for nine identified projects, among which are the following : (a) compact farm; (b) development centre; (c) irrigation; (d) seedling nurseries; (e) farm equipment; (f) integrated financing; (g) training; (h) energy plantation: and (i) marketing.

Scope of Work

The Five-Year Programme

The Five-Year Integrated Virginia Tobacco Agricultural Development Programme will cover the technical, economic, financial, organization and management and social aspects.

The technical aspect will include a study on the existing and proposed farm layout. A study on the existing and proposed production technology and the determination of the support facilities to be provided will be included.

The economic aspect will cover four phases, namely, market identification, marketing programme, projected sales, and their contribution to the national economy. The market identification study will include demand and supply analysis and determination of selling prices. The study on existing and proposed marketing programmes will include selling organization, terms of sale, channels of distribution, location of sales outlets and marketing cost. The other phases of the economic aspect will cover a study on expected volume of sales and the contribution to the national economy in terms of income, labour employed and taxes paid.

An analysis of the total programme costs, sources of financing and financial statements will be included in the financial study.

The organization and management aspect will cover a study

on the type of organization and its corresponding structure. It will include the identified roles to be performed by appropriate participating agencies. It will likewise deal with programme management systems and procedures which will include a training design and monitoring and evaluation system.

A study of the services and facilities required by the agricultural workers dependent on the tobacco industry will be undertaken in the social plan. Specifically, this will deal with educational, health and other social services requirements.

Project Action Plans

The preparation of the project action plans will also be comprehensive and integrative. These will identify activities for the implementation of the nine identified projects. Basically, the action plans will be proceeded by feasibility studies which will later be translated into operation plans. Its preparation will cover the technical, economic, financial, organizational, social and management aspects. It will also include an initial training programme for 60 community workers.

Projects Concept

Barangay Curing Service Association

The organization of BCSA is an initial step towards the social and economic uplifting of the Virginia tobacco farmers. This also manifests the farmers' unified commitment and support to the integrated Virginia tobacco development programme through their concerted effort actively to participate in the development of the industry.

This organization stems from the common need of farmers to have more efficient barns for centralized curing the produce—high quality leaf Virginia tobacco. It will be responsible for managing, operating and maintaining the curing barns.

BCSA will derive its strength from mutual co-operation of members. A set of by-laws will be formulated to establish the rights and to guarantee the benefits of each farmer-member. It will also define the duties and responsibilities expected of him.

The Association will be registered with PVTA to give it legal personality to enable it to engage in business activities. For it to be enduring and successful the association will provide not only curing services but other agricultural services to cater to the basic needs of its members.

Furthermore, the Association will serve as an effective vehicle through which assistance programmes can be charted or inputs channelled. Also, it can be an effective vehicle for the dissemination of information, education and development of capability or skills of the tobacco farmers.

Finally, BCSA will set the tone for the establishment of compact farms wherein farmer-members assume greater responsibility for their own development towards a self-reliant or self-sustaining community.

The main features of BCSA are the following :

(i) It will be a registered, non-stock, non-profit organization;

(ii) It will be a *barangay*-based organization of tobacco farmer-tillers;

(iii) It will strive to be and be guided along the lines of a community co-operative organization;

(iv) It will serve as the channel for all technical and financial assistance from both the Philippines Government and private agencies;

(v) It will be the centre of information, education and skills development;

(vi) It will set up, manage, control, operate and maintain more efficient curing barns;

(vii) It will serve as the forum in the exercise of self-government and community leadership.

Community Workers' Training

PVTA community workers' training is conceived as a measure taken by PVTA management to develop the agency's internal capabilities in pursuing its new directions. Initially it will train selected PVTA field technicians who in turn will serve as a core of trainers for other PVTA staff and for the farmers. Competencies in priority areas as required by the changing needs of PVTA will be developed in the trainees.

Although not a part of the training proper, post-training activities should form part of the over-all programme to ensure that learning acquired is applied and reinforced in the actual work situation. An evaluation and monitoring system with a self-correcting mechanism should be developed and installed to help identify other training needs and inadequacies for immediate action.

Farm Modernization Programme

To increase agricultural production, farm modernization should be adopted, especially in areas where new procedures have not been previously used, depending on the local conditions prevailing in each place. Farm modernization is a complex process and involves a long chain of innovations which may require mechanized devices that cannot be disentangled from non-mechanised devices.

Farm technology may be classified into two major categories one concerned with the use of agricultural machinery and the other with innovations in the biological and chemical sciences. Farm mechanization refers to the use of mechanical procedures or the appropriate equipment in farm operations. There are different types of mechanization involving different problems and therefore with different consequences. One type is the introduction of tractors as the power source for field operations. A second is improved pumps and power sources for lift and well irrigation. A third is improved means of transport between fields and public paths or roads and from there to the closest market centre. A fourth is improved implements for sowing, cultivation, weed and disease control, harvesting and threshing. A fifth is new methods of first stage processing of farm products.

The other category of farm technology consists of innovations derived mainly from the biological and chemical sciences, whose impact is directly through the rise in crop yields per unit of land and only indirectly in output per worker. Improved seeds, improved techniques of irrigation and crop rotation, the increasing use of fertilizer, pesticides and herbicides, new developments in the methods used in the various phases of tobacco production, all contribute to increase in farm output.

Despite the advantages of farm modernization, it does not,

however, require complete change in every farm practice, power source, farm supplier and equipment. In some cases, only one or a few aspects of this technology is ineffective; hence, it alone needs improvement to increase production.

Farm Services Programme

Along with farm modernization goes the need to establish the most efficient and suitable supporting system with the aim of directly supporting production and improving the farmer's quality of life.

This supporting system is composed of a large number of varying elements, all directed towards helping the farmer and his family to direct their activities towards one primary aim—development. These elements can be divided into three main categories, namely, farm services, such as supply, marketing, credit facilities and instruction ; various public services, such as education and health ; and physical infrastructure, including roads, warehouses etc. The structure of the supporting system must be flexible so that it can change with the changes which can be expected to take place in the economic and social structure of developing countries.

The system is made up of a number of different services whose scope, structure and nature change constantly in the course of the development process. The ability of these services to fulfil their functions is determined by three main factors : accessibility to the farmers who need them, scale and location within the rural areas.

Accessibility is largely, but not entirely, determined by the distance and time it takes a farmer to travel to reach the site of the services. But the nature of the service must also be adapted to the farmer's level of understanding.

The scale of each service determines its level and its complexity. A service unit which can provide for a larger number of clients or handle a greater quantity of goods will be better able to improve standards and provide complex, specialized activities. The scale of the units in the supporting system is directly affected by the amount of capital the national or local economy can invest in developing the system and by the ability of the client to pay a higher price for the services he requires.

The scale and complexity vary with each stage of development. As a higher stage is reached, the minimal feasible scale is also larger and the degree of complexity of the system is greater.

The location of the services determines the efficiency of the whole supporting systems. The greater the number of services concentrated in one physical location, the easier it is for the farmer to use them. Concentration of services also leads to savings in the investment necessary for creating an infrastructure common to a number of services.

These three factors determine, directly or indirectly, the efficiency and cost of the supporting system. Their relative importance changes with the process of development, as do the criteria which determine the optimum dimensions for each factor.

Discussion

The preceding case provides an overview of a project proposal addressed to a government agency desirous of making functional and systematic the type of intervention it wishes to undertake in the tobacco rural areas. Needless to mention, the approach adopted here is institutional/organizational development intervention with specific projects to be implemented in the initial phase as part and parcel of the planned intervention. Since the nature of planning is addressed to the entire organization of PVTA, it is deemed necessary that the intermediate beneficiaries, i.e. PVTA officials and field employees, will be the principal actors in the planning process. However, the development of the Five-Year Programme will be based on the close participation of the tobacco farmers, the ultimate target beneficiaries, who will be tapped during the initial phase of experimentation and piloting of packaged projects.

The focus of all *barrio*-level interventions is the Barangay Curing Service Association, around which the tobacco farmers are presently gathered. It is through this local-level organization that the joint planning teams of PVTA and DAP will work together for a realistic projection of PVTA programmes in the next five years.

Preliminary field work and research by the DAP team show a very encouraging situation where *barangay* leaders and members are enthusiastic to co-operate with this partnership. It is hoped that with the eight other projects launched simultaneously in the province of Abra the process of planning for the administration and consequently the *barangay* level will have gained action and leverage for a problem-solution approach in the true sense of the work of local-level planning.

Concluding Remarks

The current approach to local-level planning for rural development in the Philippines is indeed gaining more ground. More and more government as well as private agencies go directly into the region, province, municipality and *barrio* for more controlled and effective planning with rather than for the people. A few cases in point are the SAB—A River Basin Barrio Development Project of Leyte; the Quinlogan Agro-Forestry Project of Palawan; the Malinao Projects of Agusan, Bukidnon and Capiz. Methodologically, the approaches followed by planning agencies are as varied as there are many institutions in the country. There are, however, some unifying factors to all these planning endeavours, namely :

(i) There is growing awareness of the needs to involve the people in research, project design and implementation.

(ii) There is a realization that detailed planning becomes closely operational and more effective at the local level;

(iii) Plans become more meaningful to the people involved in the planning effort;

(iv) Local counterparts from the target areas are made readily available when they are given the opportunity to voice their own problems and suggest solutions for them, this experience provides a self-enabling process for local partners to manage their own affairs;

(v) The experience provides a mutual learning experience between the planners-implementers and the target beneficiaries;

(vi) It leads to co-operative and communal productive activities which prepare the community to take over project management;

(vii) Technical requirements in planning tend to complement existing capabilities, resources and ongoing indigenous projects, rather than bringing about drastic changes.

The following issues concerning local-level planning are

based on previous involvement of various planning activities in the Philippines. First, assuming that a local-level programme has indeed provided a given community with the capabilities to manage its own affairs, in the manner and directions it defines, what kind of delivery system must a central authority adopt in order to allow the free flow of exogenous support of projects to continue as needed by each developing community ?

Based on past experience, local-level planning has always attempted to respond to the needs at the grass-roots level. Interagency linkages as well as interregional planning efforts have often been emphasized. On the other hand, national planning has generally been industry-specific where man is viewed as an input rather than a partner and beneficiary. The mechanism that is necessary to translate national policy into detailed local plans is a subject that must be looked into.

Capability-building for implementers as well as beneficiaries of local-level plans is a long-term process which must be reconciled with the pragmatic cost of daily subsistence. The economics of experiential learning is to be reckoned with. If indeed a model is achieved by this long-term process, the question that comes to mind is : how replicable is it for cross-country implementation ?

In the final analysis, the questions that must be asked are the following : How economical is it to involve people in the process of development ? What is their share in this process other than their willingness to learn and accept the new technology brought to them by the implementers and change agents ? How detailed must the national policies be to be relevant to the local communities ?

SEVENTEEN

SRI LANKA

K.P.G.M. Perera

IN COMMON with the other countries of southern Asia, the large majority of Sri Lanka's population live in the countryside and rural landscape is predominant. The agricultural sector is characterized by the duality of (i) the plantation sector, producing crops mainly for export, and (ii) the peasant agricultural sector. There is, however, some overlap as the peasant producers also to some extent are engaged in the production of export crops on their small holdings of land.

This paper is mainly concerned with the infoımal peasant agricultural sector. The rural landscape, dominated by the peasant agricultural sector producing chiefly food crops for internal consumption, as well as some export crops on a small scale, is the area on which attention is focused. One of the most outstanding physical features of the rural landscape is the dichotomy between the irrigable low land devoted to paddy cultivation on the one hand, and the high land devoted to other crops and dominated by homesteads and village gardens.

This rural landscape and the rural economy based on it could be described as being characterized by small unit holdings of land, worked by the owners themselves, by sharecropping tenure and by hired labour. But, by far, the owner-cultivators dominate. The general level of productivity is low, mainly owing to poor methods of cultivation without sufficient inputs. The income per farmer is low because of (i) poor land productivity, and (ii) the small extent of land available to each farmer. These two facts in general characterize the rural sector and the rural economy and, at the same time, represent the key factors that lie immediately behind the unsatisfactory income situation in the rural sector.

Along with low farm incomes exist widespread underemployment and unemployment in the rural sector. In fact, both inadequate incomes and inadequate employment spring from the same set of circumstances. Given the present active population, the existing cultivable extent in a given area is inadequate which leads, over the years, to progressive fragmentation of the farms beyond economic size. Or, alternatively, given the area available for cultivation, the active farm population is excessive, leading to an increase of underemployment and unemployment. Thus, the rural sector represents an intertwined matrix of problems, each intensifying the other and feeding each other. However, if they are disaggregated the following different elements in the picture emerge :

(i) Rapid increase in population;
(ii) Negligible expansion of cultivation area;
(iii) Insufficient emigration from the rural sector;
(iv) Inadequate increase of farm output per acre.

These can be stated as the immediate causes of the present plight of the rural sector overburdened by the twin problems of inadequate incomes and inadequate employment. In fact, these factors are themselves caused by several other layers of causes.

In the above paragraphs a somewhat dismal picture has been painted, to which can be added low sanitation, poor housing, malnutrition (especially among children), inadequate medical care, poor access to education, social inequalities and a whole host of other depressing characteristics.

Is the picture really as dismal as represented above ? Is there really no way out of this situation ? An attempt is made below to answer these two important questions in turn.

The economists, government planners, foreign consultants and others, all of whom base their conclusions on quantitative statistical data (no doubt conforming to scientific requirements) take an external view of the rural economy and rural society. Facile comparisons with the cities or with conditions in more developed countries clearly appear to support such conclusions. The social and communal compensations are clearly not given any weight. But, really speaking, rural society should be viewed from the point of view of those living in that society, i.e. internally. When viewed this way, it is, however, concluded that living in the social matrix that obtains in the rural society compared to the conditions of impersonal society in the cities is generally more satisfying. It is necessary to bring out this fact, because none of the quantitative indicators associated with the external view can take this factor into consideration.

The above statement is not meant to explain away some of the very real problems of poverty, especially for the poorest sections of the rural society (even though its harshness is much less for similar levels of incomes in the cities). Is there a way set out of this situation ? Indeed there can be several different solutions possible and feasible depending on the local circumstances of each country and an attempt will be made below to set out some of the solutions adopted in Sri Lanka, which are particularly suited to the conditions in the country's context.

Successive Governments in Sri Lanka were not in a position to ignore the rural sector in any sense and the control of the elite in the cities was minimal in contrast to the recurring statements in development literature which stated the contrary. One reason for this favourable situation may be the democratic governmental system, which gives sufficient political weight to the rural society. This was reinforced by the agricultural development potential which existed in the dry zone of Sri Lanka which offers extensive possibilities of irrigation extension and settled agriculture. Each Government also looked upon the dry zone as an area of challenge, where re-building an irrigation system accomplished by the ancient Kings became a national dream and a matter of national pride. The ready public revenue that could be siphoned off the export plantation industries made possible the mobilization of financial resources for this purpose. Thus the convergence of all these factors made for a large increase of irrigable agricultural land which was, as a matter of firm policy, distributed free exclusively to the landless families of the rural sector, mainly from the thickly populated wet zone areas. It should be restated for emphasis that the main beneficiaries of these massive investments were the poorest sections of the rural society. This policy is being continued at an unparallelled pace, and landless families will continue to benefit from this programme.

Similar support was also planned for by institutional changes Since a large proportion of rice lands, which is the main productive resource of the peasants, is traditionally cultivated under the sharecropping tenure system, elaborate legislation was promulgated and implementing machinery set in motion to ensure permanent tenure for the sharecropper and to regulate the rent payable to the landowners. In fact a revolutionary concept of two distinct ownership rights—(a) a cultivation right to the sharecropper, and (b) ownership rights to the landowner, both rights to be sold or inherited separately —was brought into force.

Expansion of institutional credit for agricultural production purpose was expanded, to relieve in particular the small farmers from the high interest rates levied by the moneylenders.

In addition, successive Governments have adopted national programmes to help the rural society, such as a free education system and extension of schools to the rural areas, a free health care system and extension of hospital facilities to rural areas, guaranteed price schemes, and agricultural extension services, in addition to purely distributive measures such as subsidised food schemes, from which the rural poor draw substantial benefit.

Sri Lanka had been conscious of the need for national planning as an aid to economic development (although in actual practice the performance of the economy has been uneven and not commensurate with expectations, partly and perhaps substantially owing to unfavourable external factors) and had established elaborate national planning machineries. Particularly from 1965 onwards, arrangements were evolved so that a "planning" input entered into all major economic decisions of the country, and in the effort to guide the economy, elaborate systems of controls, incentives, subsidies, taxes, duties, investment guidelines and massive government inflows of investments into desired sectors were carried out. In all formal economic plans, as well as policy statements, and in the actual public sector investments, the rural sector was given an important place. Thus there were very strong national programmes which heavily supported the rural sector.

Colonization Schemes

Flowing from some of these activities, large areas of certain river basins in the dry zone were developed virtually from a state of virgin jungle to prosperous agricultural settlements, with all the social and economic infrastructure. In point of fact, they can be described correctly as area development or regional development schemes with a heavy input of physical planning. As stated before, this activity is proceeding at an ever increasing pace with very strong support from the international community as well. This exercise has all the elements of integrated and comprehensive multisectoral development. Thus, within the scope of definition of subnational area planning, this category of development is the first in historical sequence. Such areas are called *colonization schemes* in Sri Lanka.

As mentioned earlier, the dry zone of northern and eastern Sri Lanka was covered with an extensive network of large and intricate irrigation structures supporting a prosperous agricultural economy. In the course of historical time, the original structures went into decay as a result of external invasions, internal strife and the advent of malaria, and almost all such areas returned to jungle and forest. In the 1930s repair and restoration of these works were commenced and "colonists" from the rest of the country were settled. Up to and since Independence in 1948 this activity has continued. Since 1948 a whole new river basin (Gal Oya) has been studied and an irrigation-based development programme for this entire basin was comprehensively planned and implemented. Water resources were developed for multipurpose irrigation and power development, paddy tracts, channels, homesteads, roads, and schools were planned and laid out.

Similar schemes followed in river basins such as Kala Oya and Walawe Ganga and elsewhere. The most recent and the most stupendous scheme now undertaken is the massive Mahaweli Ganga Development Project which involves several trans basin diversions as well.

The main point of interest in all these schemes in the present context is that they are very much integrated area development schemes, and that as a consequence of this activity by the Government, prosperous and self-reliant small-farmer communities have been built up. In short, these farmers are assured supplies of water, own 3-5 acre holdings of rice land, produce high yields, have a substantial surplus for sale and enjoy high incomes. To achieve a condition such as this, is clearly the aim of rural development efforts.

But has the country enough land and investment resources to be able to provide such a cultivation unit to all the farmers in the rural sector ? Clearly not. Therefore, other solutions will have to be found, as practical means to bring about the development of the rural sector.

Divisional Development Council Scheme

In 1970 the Ministry of Planning and Economic Affairs established a separate division, called the Regional Development Division to work out an approach complementary to the national and aggregative planning, which was provided with a large staff as well as financial resources for capital expenditure. The basic philosophy of this programme was to introduce local planning over restricted areas so that local resources, skills and knowledge could be harnessed locally for the economic development of the area itself. The rationale is completely justifiable because national planners are concerned primarily with the main national economic aggregates and the view from such a high level would miss the local resources and possibilities for development. This would also enable enormously large numbers to be drawn into the planning and development activities. Doubtless this was an innovative and ambitious approach to decentralizing planning activity as well as reducing the area unit of planning to a micro level. The Five-Year Plan explicitly stated that through the District Development Committees and the Divisional Development Councils it was proposed to decentralize the work of plan formulation. The Divisional Development Councils (DDCs) were expected to be the main link between the network of government agencies on the one hand and the local community on the other.

In 1970 and 1971, DDCs were set up for each local authority area, each Council consisting of representatives of people's organizations such as the village committee, rural development society, cultivation committee, women's *samities*, co-operative society, and also of divisional-level officers. The chairman of each DDC was the Divisional Revenue Officer (since renamed Assistant Government Agent) and its secretary was a Divisional Development Assistant (a cadre of graduates recruited to the Regional Development Division with these activities in view).

The Divisional Development Council was expected to meet once a month and identify local resources, potentialities for development and development projects ; the people were assumed to participate through their representatives in a big way in the planning exercise, and local resources and skills were to play a dominant role in the activities that followed. Adoption of local technology, improvement of local skills and techniques and the use of appropriate technology were key strands of thought underlying the programme. Co-operative ownership and management was another.

Project formulation was done by the planning staff at the DDC level, approved at the district office and forwarded to the head office for final approval and funding. Management was by the existing multipurpose co-operative societies and, where such organizations were not able or willing to manage these projects, special co-operatives were formed for the purpose. A substantial part of the funds was given as grants and the balance at very low rates of interest and easy repayment terms through the branches of the People's Bank.

In 1975, the DDCs were reorganized and the area of operation became the electorate. As a result the 607 DDCs were reduced to 145. The membership was reduced to 15, consisting of 5 officials and 10 others and the Member of Parliament became the chairman of the DDC. A DDC secretariat was established as an independent unit, to which the Development Assistants were attached.

Along with this change, elaborate development co-operative societies were formed to manage the DDC projects.

A total of around SRs 135 million has been released by the Regional Development Division in respect of about 2,600 projects, and the number reported employed at the end of each year was as follows :

1972	...	6,398
1973	...	13,872
1974	...	18,374

1975	...	20,457
1976	...	33,271
1977	...	19,758

The experience of most projects was that they were basically not viable, mostly because of poor management, and were suffering losses except in a few cases. For them to go on, continuous infusions of funds would have been necessary, a procedure not justified in comparison with the benefits.

Seeing the programme in perspective after nearly seven years since the inception, the enormous initial enthusiasm for this programme on the part of government officials, the public and the Members of Parliament should first be noted. But the cardinal mistake of this programme was that its attention was concentrated solely on starting projects under its own management ; it paid little attention to the ongoing economic activities of each area. However, if the Divisional Development Assistants, who joined the main cadre of the programme, had focused attention on the existing economic activities of their respective areas, studied them, identified problems and attempted to improve those activities, better results would have been achieved and the effort would have been more appreciated by the people, who would have become more willing and co-operative participants (as well as beneficiaries) of any extended planning exercise.

The current Government has taken stock of this programme and its poor performance so far and has decided against investing any more capital expenditure in it. The Government does not in any way desire any existing and viable projects to be closed down, but since most projects are extremely poor in management, many of them will in course of time cease to function. Further, some of the projects are being taken over by relevant government departments such as the Small Industries Department.

Even though the Government has decided against proceeding with the DDC programme as originally envisaged, this does not mean that the local-level planning approach is not being pursued. In fact, the contrary is true, as will be seen in the following section, which describes the district planning activities which are currently being pursued with great enthusiasm and interest.

The District Level

In the area of subnational planning, the Government in 1976 initiated discussions with international financing agencies with a view to securing assistance for integrated rural development projects for administrative districts.

There are 22 administrative districts in Sri Lanka (to be increased to 24 districts shortly so that each district would on an average have about 600,000 people and an area of about 1,000 square miles, although there are wide variations between districts.

The rationale for this approach, in common with any other area-based subnational programme, is the need to supplement aggregative national plans and large national-scale projects with a more detailed examination of the resources and problems and potential to be carried out over more limited areas, so that an investment programme more specifically tailored to the particular needs of each district could be evolved and implemented.

Another fact that adds to the need and relevance of district development planning is the increasing trend towards decentralized administration and decision-making observable from about 1970. Several institutional changes that were made to give effect to this thinking include the introduction of Divisional Development Councils in 1970, the decentralized budget and the District Political Authority in 1973, culminating in an important constitutional change with provision for the appointment of district Ministers (the first district Ministers are to be appointed very soon).

The importance of the district as a subnational decision-making unit will be further emphasized when the funds under the decentralized budget are allocated on a district basis (in contrast to the present electorate basis), and even more important is the fact that, under the new Constitution, Members of Parliament will be elected not for each electorate but for a district as a whole.

At present there are three districts for which integrated

rural development planning has been undertaken (and several more will be taken up shortly), namely ;

(a) Kurunegala District (area 1,842 square miles, population 1.2 million);

(b) Matara District (area 481 square miles, population 600,000) ;

(c) Hambantota District (area 1,001 square miles, population 400,000).

Some details of each planning exercise are given below.

Kurunegala District Rural Development Project

A project plan has been drawn up by a group of United Nations advisers along with local officials. All the important sectors have been intensively studied in this district and these have indicated the critical bottle-necks in each sector and specific proposals in terms of investments and institutional improvements have been formulated to overcome them and enable the economy of the area to realize rapid growth and consequently the people to realize a higher standard of living and incomes.

The district has a land area of 1,842 square miles and a population of about 1.2 million. The total investment package under the project amounts to about SRs 400 million and a project implementation period of five years is envisaged. The World Bank has reviewed the project and has indicated tentatively the possibility of long-term IDA credit to cover about 70 per cent of the total costs.

Some of the main proposals in the investment package are :

(i) Rehabilitation of all the major irrigation works, two medium irrigation works and about 500 minor irrigation works so as to provide irrigation water to the extensive paddy lands;

(ii) The physical improvements necessary for the adoption of better water-management practices in paddy cultivation;

(iii) Provision of fertilizer stores, transport facilities and plant nurseries required for the improvement of coconut cultivation;

(iv) Provision of fertilizer stores, transport facilities, nurseries and demonstration plots for crops to be grown under the coconut;

(v) Supply of tractors for paddy cultivation;

(vi) The provision of essential requirements in the health care system to strengthen the existing services;

(vii) The provision of essential requirements in the educational system to strengthen the existing services;

(viii) Improvement of certain roads;

(ix) Construction of 10 rural electrification schemes;

(x) Provision of a large number of drinking-water wells in selected areas.

Some of the institutional improvements and innovations are :

(i) Introduction of a credit scheme for coconut fertilizer application;

(ii) Introduction of a credit scheme to enable farmers to purchase tractors for farm use;

(iii) Adoption of an intensive agricultural extension system.

This is the first district planning exercise in Sri Lanka and its implementation is expected to have commenced by early 1979.

Matara District Rural Development Project

Matara District in southern Sri Lanka has an area of 481 square miles and a population of about 600,000 entirely located in the wet zone of the country, which is densely populated. This area offers wide scope for the improvement of paddy cultivation, tea and coconut production and fisheries development. Improvement of drainage conditions can upgrade the paddy land, reduce the risks of crop damage and thereby offer more favourable conditions for improved cultivation practices.

Sweden has assisted in the preparation of the plan and will also support its implementation. Here, too, sectoral studies are being undertaken and at the same time critical investment needs are being identified. Even though all the studies are not complete, the implementation of certain clearly high-priority proposals is expected to commence early in 1979.

Hambantota District Rural Development Project

This district adjoins the Matara District and is located in the southernmost part of the country. It has an area of 1,000 square miles and a population of about 400,000. This project is assisted by the Norwegian Government and the approach is similar to that of the Matara Project. Over large parts of this district the climate is very dry and irrigation improvement and extension is a prerequisite for raising agricultural productivity. There are also wide possibilities for fisheries development. Certain high priority projects are to commence implementation from early 1979, concurrently with the detailed planning exercise which will proceed alongside. It is also planned to take up a few more districts in the near future, for study and plan formulation.

In each case, the common approach includes a detailed study of the district economy and its many sectoral activities, generation of information, identification of bottle-necks and problems and the realizable potential within the medium-term time framework, and the identification of realistic development proposals that can be implemented within the available financial framework. In contrast to the national programmes, each plan represents a distinct multisectoral package of area-specific investment proposals and institutional improvements suited to the district.

All the above three districts have predominantly rural economies and as such the development plans have many features common to area development as well as to rural development. Conceptually, these two approaches could be distinguished from each other, yet in actual practice the two approaches overlap considerably and may even become a matter of nomenclature rather than substance.

These district programmes have been described as integrated development efforts. As is well known, there is no universally accepted definition of integration in this context, and there are several equally valid connotations applicable to the term. Further, integration is not an all-or-nothing phenomenon, as there can be any number of degrees of integration either within the sector or within an area or between sectors and between areas. Commonly, there can be much disagreement as to whether any of these programmes are sufficiently integrated to justify the use of this term to describe them. One important observation that can be made in this context is that all these districts (except for a small part of Hambantota District) have hardly any scope for expansion into new settlements and consequently merely propose certain activities and projects on the existing rural landscape which is already to a very great extent integrated. Thus most of the proposals have the function of eliminating bottle-necks, or filling an important gap, or providing for an important need, and consequently the proposals appear to be a disjointed collection of projects and activities, which are not integrated with each other. There is no way of avoiding this misleading picture, because the proposals are individually integrated with the existing characteristics of that sector or area.

In selecting particular districts for this planning and development activity which consequently would result in an increased flow of technical and financial resources into the district, due care has been taken to see that districts which already have large investment flows, on account of large national-scale projects being located there, are not selected.

Replicability is also an important consideration in the district planning approach. Financial commitment which limits the scope, magnitude and framework of the proposals included in the district plans is such that this exercise could in the course of a reasonable time frame be replicated in other districts as well. But just the same, the magnitude of the proposals should be and is above the minimum threshold size, so as to make a considerable impact on the income and employment levels in the area. The district, in view of its still large size in terms of both population and area, necessarily retains many features of top-down aggregative planning, though the region's specific resource endowments are the basis of the planning exercise. In view of the above, one very important question that should be raised is whether district planning as a complement to national aggregative planning exhausts the possibilities or completely or adequately meets the needs of multilevel planning in the country's context. The limited experience in Sri Lanka in this area, however, suggests the need for another level of planning—complementary to district planning. A natural unit for this exercise would be the electorate (average population 85,000). An alternative unit would be the Assistant Government Agent's Division, which is somewhat smaller.

The district-level programme would concentrate basically

on the provision of facilities and opportunities for increased production by the Government and its agencies, such as better irrigation, roads, electrification, agricultural extension, credit inputs and marketing facilities, which may most economically be planned over a larger area such as a district.

The electorate-level programme would concentrate basically on the increased use of the facilities provided. The bulk of this activity would not involve heavy investment of capital resources but revolve more on the organization of economic agents and participants, mainly the small farmers and others of the informal rural sector whom the rural development approach is designed to serve.

The Electorate Level

Reference was made earlier to the Divisional Development Councils programme which commenced in 1971, which was an attempt at local-level planning, but which in fact turned its back on integrated development and concentrated on starting and managing a multitude of individual projects unrelated to each other. The DDC programme originally covered the area of a local authority (namely, the village committee area) but the DDC never functioned in the way it was originally formulated and the Development Co-operative Society became the institution that the DDC programme built up. The area of operation of the Development Co-operative Society was the electorate.

The Development Co-operative Society is not clearly suited for any local development planning and therefore the building of an apparatus geared for this activity has to commence right from the start. The Ministry of Plan Implementation has realized the need for this level of rural planning, but since there was no valid experience in this field, a very cautious approach had to be adopted on an experimental basis which is described below.

The *Katugampola electorate*, comprising 160 villages located in the Kurunegala District, was selected for this exercise. This is a rural area typically dominated by coconut and paddy cultivation. Both crops offer wide scope for improvement in productivity by better cultivation and management practices. In addition, there are extremely wide possibilities of a second crop or a crop combination under coconut selected from pineapples, bananas, pepper, coffee, turmeric, ginger, cocoa and vegetables. All these activities would not only enhance productivity and income but, equally important, increase employment potential at the same time. The main aim of integrated rural development planning at any level is to bring about a rapid increase in rural incomes, production, employment and the general standard of living of the area and the people by making certain investments as well as by the introduction of organizational and institutional improvements. In contrast to national or district planning exercises, in local planning the aim is to operate at the village or grass-roots level where the problems actually exist and where the solutions, too, have to be worked out.

In order to achieve the aims outlined above, the following activities are being carried out :

(i) It is necessary to take a close view at the problem structure of the area and evolve a possible solution structure in close co-operation with the community, political leadership, and the district officials (including planning staff). One method employed was regular meetings of these different groups for exchange of ideas. The second was a village survey which has been carried out over most villages. This also helps to publicize the project and arouse enthusiasm;

(ii) An increase in the efficiency and effectiveness of existing government service organizations, namely, those related to production sectors, such as agricultural extension, credit services, marketing services, input supply services. To achieve this, the programmes of each government agency for the electorate are discussed and the progress and the problems reviewed at the electorate meetings.

The co-ordination of all activities in connexion with this project is done at the office of the Assistant Government Agent at Pannala, and by the Member of Parliament, the Assistant Government Agent and the Planning Officer with the active support of the Government Agent at Kurunegala, the district centre. An extra-administrative organization has been established, based on the village development committee at the lowest level, consisting of 10-20 members drawn from the Rural Development Society and other voluntary bodies in the village. About 150 such village development committees are finally envisaged.

A number of functions are expected from these committees, namely :

(i) Identification of problems and development possibilities of the village area;

(ii) Education of the village in respect of the above possibilities;

(iii) Organization of the village for collective action, such as is required in paddy cultivation or in organizing other agricultural activities;

(iv) Obtaining the co-operation of local and divisional officials in getting more effective service from them;

(v) Functioning as a key development-oriented unit in general in the village.

Each of the village development committees will send a representative to the Electorate Development Committee set up at the Development Office. These representatives also form development sub-committees to deal with the following :

(i) Agricultural development;
(ii) Livestock development;
(iii) Irrigation and roads;
(iv) Lands and plantation crops;
(v) Electricity, construction and small industries;
(vi) Credit facilities;
(vii) Transport, posts and telecommunications;
(viii) Education;
(ix) Health and community services.

The chairmen of these sub-committees necessarily serve on the Electorate Development Committee, which is the main steering and co-ordinating committee for the project.

The Electorate Development Committee is chaired by the Member of Parliament and is composed of the key divisional officers as well as chairmen of the sub-committees and some officials from the Kachcheri headquarters.

This electorate development programme is certainly an effort of an experimental nature, and as such the above arrangements are based on conception of the needs of the organizational framework rather than on any specific experience and therefore will need much revision as the programme develops. It is very clear that it is near impossible to institutionalize such a system and expect it to work satisfactorily. In the case of the Katugampola development project, the main driving force has been the Member of Parliament, who devotes a great deal of time and effort to this exercise.

Since the programme has now been going on for some months, experience has demonstrated that this is indeed a very fruitful approach. Since the problems that this programme wishes to identify are small-scale village-level ones, their solutions and the resulting benefits could take place within a short time. This is certainly one important aspect.

One significant improvement that could be mentioned here is that the different government service agencies in the electorate have been able to perform a better service than before. Similarly, some new activities have been established, e.g. a large scientific plant nursery, with a view to supplying the needs of the electorate from the funds released by the Ministry of Plan Implementation to the Development Co-operative Society. This Society will soon be provided with two trucks, which will be hired out for various activities as well as used for the transport of plants from the nursery to the villages. It is hoped to organize an extensive planting campaign in the villages during the next rainy season. In organizing the villages for this activity, and getting as many as possible involved, each village development committee will play a crucial role.

Another good example of the type of activity that can be undertaken by a programme such as this is the provision of high-quality seed paddy within the electorate itself. It is a common occurrence that the farmers find that they are unable to obtain sufficient good quality seed paddy of the improved varieties, since the demand is concentrated within a few weeks. This problem was brought up and discussed and the Department of Agriculture was able to organize the production of a substantial quantity in selected farmers' private fields to be purchased under the Government's guaranteed price scheme.

Particularly in the case of paddy cultivation, there is much scope for improvement by means of following a common cultivation calendar in any paddy tract. One of the activities envisaged for the village development committees is to organize the farmers before the cultivation season commences, and agree on a common cultivation calendar. This will enable the farmers to secure the inputs collectively. A further, very real benefit is the possibility of better water management that this practice would permit. This type of group action has to be organized in every village and here is one instance when popular participation matters for the success of the programme.

The Government has given high priority to reorganizing the rural development society for every village (numbering about 20,000 in all). This voluntary village organization is expected to play a key role in the economic upliftment of the village in addition to its traditional role as a people's body devoted to community and social activities. When these societies are firmly established, they could be a fresh source of strength to village development and to the local-level planning exercises aimed at village development.

The Village Level

Even though the electorate is the area over which this level of integrated planning is to take effect, it is in fact the *individual village* which is the unit building block. Therefore the experience obtained from individual village development exercises become immediately relevant. There are many examples of such pilot village development in Sri Lanka, such as : (a) the Beminiwatta field laboratory area, by the Agrarian Research and Training Institute (ARTI), (b) Narangoda village by the Marga Institute, and (c) Halpe village by the Rural Development Society.

In the case of Beminiwatta (Kegalla District), ARTI organized the activities, namely, group cultivation of paddy and a group scheme for poultry and banana cultivation. The paddy farmers were organized to adopt a common cultivation calendar and improved cultivation and management methods. Input supplies and institutional credit were arranged. Poultry rearing and banana cultivation were similarly organized. Within a short period the results were remarkable in terms of increased income and employment.

The Marga Institute has conducted detailed surveys of the 250 households and their resources in Narangoda village, located in the Katugampola clectorate, organized the villagers into different groups and commenced several development activities with special emphasis on raising agricultural incomes. According to the survey data, a clear possibility of raising village incomes by 50 per cent exists over a period of five years. This indicates the general scope for improvement in the case of other villages.

Halpe village was a case where the introduction and widespread cultivation of sugarcane transformed the economy of the village, resulting in higher incomes. The value of these and other similar studies lies in indicating the potentialities of development available in average villages and possible organizational approaches that can be adopted in a more general way. In an electorate planning and development exercise, all this experience becomes immediately relevant because the individual development unit is still the village itself.

A commonly observed phenomenon in most cases of externally induced village development such as those mentioned above, is that very soon after the external support and assistance is withdrawn, many of the innovative activities cease to lose their intensity, and the more traditional level of activities returns. In short, externally induced momentum does not appear to become part of the village "system" and perpetuate itself leading to self-sustained change and improvement.

Conclusion

For about a decade, increasing awareness of the inadequacy of national aggregative planning by itself has been expressed and the need for a closer ground-level approach with emphasis on more area-specific local-level planning has been realized. Many countries have accepted this position and have embarked on various local-level planning approaches, and these efforts have generated a certain amount of experience under diverse local and national conditions. The main aim of this paper has been to describe some Sri Lanka's experiences in this connexion. It would be valuable to pool the diverse experiences, successes and failures to enrich understanding of the factors that underlie the processes of rural change and improvement, with regard to which our concept and knowledge are far from complete.

The basic aim of integrated rural development programmes

is to increase incomes, employment and the general well-being of the country and the rural people, while at the same time ensuring that the broad mass of the rural people, especially the rural poor, are drawn into this planned process of change as active participants and beneficiaries.

In order to set this process in motion, the first requisite is a deep understanding of the dynamics of economic and social change in the rural context. Each rural society is set in a particular social, political, economic and cultural milieu, so that there is no assurance that uniform prescriptions can be worked out that will have widespread applicability. The need for understanding of each society is clear.

Beyond that is the need for deciding on a set of measures and an organizational framework to translate them into practice. This needs a conceptually separate series of studies for each area and each sector within the rural universe. These studies take time and need suitably qualified personnel to undertake them. The shortage of such personnel, in countries such as Sri Lanka, has necessitated obtaining the services of expatriate experts.

The other aspect with regard to formulation of integrated rural development programmes is the time and effort required to formulate them. A compromise that has been adopted in Sri Lanka is to proceed with the planning exercise, along with implementation of some of the clearly high priority projects. This, however, exposes the exercises to the common charge that the programme is a collection of *ad hoc* unintegrated projects.

There has been strong consensus among the advocates of integrated rural development that the interests of the rural poor should receive precedence, and they should be intimately involved in all stages of planning and implementation, and indeed this is stated as a prerequisite for its success. Yet in actual practice how this particular segment of rural society can be made to participate, or what particular arrangements would make this practicable, are not very clear.

There remains the further issue of the scope of changes that can be proposed and incorporated in an integrated rural development programme. The ideal pre-conditions of planning would be a strong political commitment, a suitable administrative structure, minimum disparities in income, land ownership, social status within the rural society and the existence of a high potential for economic advancement in terms of resources and skills and attitudes. These ideal conditions cannot be expected to be found in actual practice and what is to be found in reality is a set of circumstances which gives a very restricted freedom to the scope of any plan of action.

Another important issue is how far each development programme would make the desired processes of change become self-sustaining.

Finally, there is the question of the areal unit over which integrated rural development should be planned in order to realize the best possible gains. Should it be some natural region or economic region, or a district or an electorate or a smaller area of a few villages, or the individual village itself? Each has its own advantages in a given situation and the correct answer may be a combination of multilevel planning, suited to a particular context.

These and many other vaild issues can be raised, some theoretically interesting and intellectually stimulating, others more operationally relevant and therefore of greater concern to the practitioners. These make integrated rural development an area of challenge, with the as yet many gaps in our understanding to be made good, it is to be hoped, with increasing experience in many diverse situations. Every empirical support is indeed necessary to refine and benefit from integrated rural development, which is essentially a pragmatic concept and approach. If empirical and operational content is not rapidly forthcoming, integrated rural development can deteriorate from being a promising theoretical construct into a catchword.

EIGHTEEN

THAILAND

Vithya Siripongse

Introduction

THAILAND, for the past two decades, has accepted development planning as a tool to accelerate economic progress with one of the most cherished hopes that poverty would be contained and eventually eradicated. Thus, the First and Second Plans, covering the periods 1961-1966 and 1967-1971 respectively, were promulgated and accepted as national development objectives aiming to rationalize better public resources allocations as well as to promote vigorous private investments.

Under the two Plans, development policies focused primarily upon economic and social development at the national level in order to lay down the foundation for the future growth of the economy. Thus public sector investment objectives concentrated on the construction and rehabilitation of basic national infrastructure facilities, such as highways, electricity and irrigation (superstructure) systems, schools and hospitals. The resultant economic growth during the 1960s was quite remarkable and can undoubtedly be attributed in part to the successful achievements of the first two Plans and, in particular, to the rapid increase in public investments as well as to the relatively more favourable international economic conditions. During the period, the gross domestic product, at constant prices, grew at an average rate of 8.1 per cent annually and *per capita* income almost doubled.

However, the overriding issues inherited from the economic growth of the 1960s are the ever increasing regional gap in economic welfare and the disparity of incomes among different occupational groups, which are becoming more pronounced, especially among the rural population. These problems had become more explicitly the concern of policy-makers and planners alike and had received increasing attention during the Third Plan (1972-1976). Serious efforts for a more effective analysis and planning for development at the regional level were undertaken in the north-east and the north with technical assistance from foreign sources. This challenging and difficult task, however, was not yet expressly comprehended by national-level planners and policy-makers at the time. Thus, these important undertakings were carried out on an *ad hoc* basis and development activities that followed were not well supported nor spatially co-ordinated to accelerate satisfactorily the pace of economic and social development of these lagging regions.

Not until the formulation of the Fourth Plan (1977-1981) development strategies did the critical issue of "distributive" strategy become an adopted national development policy objective, when clear-cut development guidelines, aiming to reduce the existing socio-economic disparities and accelerate rural income and employment opportunity, were methodically conceived and advocated in the national Plan and are now being seriously pursued.

Conceptual Overview

In accordance with the Fourth Plan strategies, local development planning in Thailand has added another important dimension of the bottom-up concept of planning "with" the people to the already established top-down process "for" the people. This concept has three distinct socio-economic objectives, namely :

(a) To meet basic felt needs in the predominantly rain-fed rural areas where, for the past two decades, the people have been denied the fruits of development;

(b) To provide adequate rural infrastructure and related productive inputs in order that they will be in a better position to help themselves;

(c) To enhance further, on a step-by-step basis, the capability of local-self-governing bodies to become more responsive and viable tools for integrated development in the rural areas.

With this conceptual framework in mind, the so-called Provincial Development Plan has emerged as a desirable bottom-up process within the context of the Fourth Plan development strategies.

Basic to the above objectives is the fact that the magnitude of rural problems and the diversity of felt needs of the rural people, which are far too numerous for the absorptive capacity of central implementing agencies, can no longer be met by the traditional top-down process alone nor be kept waiting for the fruits of such development. Certain action-programmes must be initiated to alleviate them and to satisfy the rising expectations of the rural people. Furthermore, it is hoped that relevant data and information generated from the local-level planning process would constitute the most valuable feedback into the regional and national planning processes. Thus, local-level planning can initially be looked upon as an important supplemental development measure; while in a longer perspective, if such efforts prove successful and necessary professional expertise gradually develop, it can then be recognized as an accepted bottom-up process.

In order to facilitate a systematic approach towards decentralization of planning to the provincial level, the Government promulgated a provincial planning regulation in July 1977, stipulating that the authorities of every province, except the Bangkok Metropolitan Authority, must formulate a five-year development plan under which careful analysis of existing over-all socio-economic conditions and problems must be undertaken and the basic felt needs of the people, together with specific project and programme proposals as well as financial input requirements of each province, must be fully reflected. The regulation further envisages a two-level development planning mechanism, each of which has specific function and responsibility as well as authority.

The Central Committee for Provincial Development (CCPD), chaired by the Minister of Interior, is composed of some 17 members at the Under-Secretary of State and Director-General levels from key ministries. The secretariat of CCPD comes under the responsibility of the Director of the Regional Planning Division, National Economic and Social Development Board (NESDB). CCPD is vested, principally, with the co-ordination and integration of provincial planning and development activities with respect to regional and national objectives and, more importantly, with the power to allocate needed financial resources for provincial plan implementation.

The Provincial Planning Committee (PCC), chaired by the provincial Governor, is composed of some 20-25 departmental representatives of field level and representatives of local self-governing bodies. The secretariat of the PPC comes under the responsibility of the Chief of the Provincial Planning Unit, attached to the office of the provincial Governor. PPC is vested with the development planning function and has responsibility to co-ordinate and integrate both the topdown inputs and the bottom-up basic felt needs of villages, communes and districts. Furthermore, each year PPC has to prepare an Annual Operation Programme in close co-operation with the Regional Planning Centres of NESDB, for submission to CCPD for review, approval and budgeting prior to implementation.

In summary, this bottom-up planning and development process, as briefly outlined above, aims primarily to rally more actively popular participation and the self-help spirit of the rural people in the formulation and especially in the execution of the much-needed small and simple projects and programmes which can be best improved and implemented as well as operated and maintained by the people themselves. From Thailand's experience, it is equally important to stress at this juncture that this bottom-up process unmistakably requires clear-cut policy directives as well as sufficiently strong institutional and especially financial supports from the Government. Thus, in the fiscal year 1979, for the first time ever the Government has specifically earmarked 1 per cent of the national budget for provincial plan implementation purposes.

Practical Issues in Local-Level Planning

The following issues are raised as a basis for further discussion.

The central problem facing development planners in a developing country such as Thailand is the question of the suitable methodology to be employed. The air-conditioned "ivory tower" technocrats tend to view local-level development planning from the highest plateau of technical perfection, while the "folksy" outdoor practitioners view it simply from the plain of an evolutionary process that needs up-hill cumulative experience-building over time.

The adopted conceptual framework and methodological approaches towards local-level planning and development in Thailand are somewhat more specific in criteria and objectives, and most simplified in the planning process and analytical techniques employed. This is due not only to serious local constraints with respect to existing socio-economic conditions, time factor availability, and the paucity of required basic data and qualified personnel involved, but also to a theoretical conviction concerning the very nature of local-level planning itself. It is believed that in a developing country such as Thailand effective planning should primarily be a creative response to a particular set of prevailing circumstances which challenge the decisions of the policy-makers. These are namely, socio-economic and political commitments to the people, constraints of the existing institutional framework at national and local levels, absorptive capability of local professionals and, in particular, the rising expectations of the people, all of which dictate choices of planning criteria and methodology to be employed. Under the existing conditions, it might not be altogether wrong to say that, in reality, planners in developing countries seldom enjoy the luxury of time and choices of methodology usually accorded to residents of the ivory towers if immediate action programmes are to be realized. It is almost a matter of a "now or never" decision.

Closely related to the above issue is how "comprehensive" local-level planning should be. This semantic question has been exhaustively debated among those concerned and a compromised solution reached ; that the comprehensive model advocated by the United Nations Centre for Regional Development would be adopted for provincial planning during the next Five-Year Plan period, prior to which intensive and thorough training programmes have to be initiated in order to equip provincial planning officials with essential analytical know-how of the multifaceted aspects of a comprehensive and integrated plan ; while the current five-year provincial plan, in its existing form, should be selectively modified and improved to meet the so-called standard format, a bare minimum requirement of a provincial plan within the context of the existing guidelines.

Another important issue is a somewhat more down-to-earth question of how "local" is local-level planning. It is indeed very difficult, if not always impossible, to come up with a universal solution or even a recognizable definition for it. However, it might at best be defined within the context of national as well as regional development strategies and objectives which largely determine the extent and direction of local level planning and development in the short and medium terms. Nevertheless, it might be interesting to note that, under the existing administrative organizations at the provincial level, Thailand has 71 provinces (excluding the Bangkok Metropolitan Area), with approximately 600 districts and subdistricts, more than 4,500 communes, and about 50,000 villages. Also, under the provincial administrative system, there exist local self-governing bodies which are spatially grouped for administrative purposes into Municipality for urban areas, Sanitary District for semi-rural areas, and *Changwat* (Province) Administration Organization for the predominantly rural areas. Thus, the mere thought among energetic and ambitious bureaucrats of decentralization of the planning process further down the provincial administrative hierarchy would indeed become a nightmare for Thai planners.

Decentralization of planning and development activities poses another critical issue of proper interdependency mix between devolution of authority vis-a-vis centralized co-ordination-cum-financial allocation authority in order to achieve better harmony in development activities. Local authorities yearn for complete decentralization of both responsibility and authority. They often claim that if effective local-level planning and development were to become an ultimate objective, they can best be achieved by granting more autonomy to them. However the varying degrees of interdependency mix and proper function-sharing between the central and local governments under the existing institutional framework may differ from one developing country to another, the issue still remains and is becoming more critical but less academic than ever.

The last but not the least issue has to do with international assistance programmes (bilateral or multilateral). These have not yet been fully geared towards supporting bottom-up planning and development for integrated rural developing countries. On the technical assistance side, more diversified and less rigid training programmes, adapted to meet practical short-term as well as more graduated long-term requirements of local professionals, could be provided. On the capital assistance side, the usual economic efficiency yardsticks and associated pre-conditions, formerly applied to measure feasibilities and viability of national and regional projects and programmes, may have to be forsaken if planned development at the local level is to be further accelerated.

Conclusion

The brief account of the development of local-level planning activities in Thailand and its implications, though relatively recent in experience, are by all means very fundamental. Were these practical issues to be specifically and systematically finalized, they might lead to more meaningful integrated development efforts.

SELECT BIBLIOGRAPHY ON RURAL DEVELOPMENT AND LOCAL LEVEL PLANNING

Books, Monographs and Occasional Papers

Agricultural Projects Services Centre (APROSC), Kathmandu, Report of seminar/workshop on peoples' participation in rural development in Nepal. Organized by the Ministry of Home Panchayat and APROSC, Kathmandu, 24-26 July 1978. Kathmandu, APROSC, 1978. 104 p.

Almario, Emelina, et al. *National profile of poverty in the rural Philippines.* Quezon City, Institute of Philippine Culture, 1976. iv. Biblio.

Aqua, Ronald, *Local government and rural development in South Korea.* Ithaca, N.Y., Cornell University, Center for International Studies, Rural Development Committee, 1974. 69 p. (Cornell University, Center for International Studies, Rural Development Committee. Rural Local Government no. 13), Biblio.

Aqua, Ronald, *Local institutions and rural development in Japan.* Ithaca, N.Y., Cornell University, Center for International Studies, Rural Development Committee, 1974. 102 p. (Cornell University, Center for International Studies, Rural Development Committee, Rural Local Government no. 8), Biblio.

Asian and Pacific Regional Agricultural Credit Association, Bangkok, "Report on the first general assembly of the Asian and Pacific regional agricultural credit association and third Asian conference on agricultural credit and co-operatives (ACACC), 10-14 October, 1977", New Delhi. Bangkok, APRACA, 1978. iv.

Asian Centre for Development Administration, Kuala Lumpur, *Approaches to rural development in Asia.* Kuala Lumpur, ACDA, 1975 3 v.

Asian Centre for Development Administration, Kuala Lumpur, "Strategies of rural development in Asia : A discussion". Kuala Lumpur, ACDA, 1976. 160 p.

Asian Development Bank, Manila, *Rural Asia*, Singapore, Federal Publications, 1977. 489 p.

Axinn, George H., *New strategies for rural development*. DeWitt, Mich., Rural Life Associates, 1978. 194 p.

Aziz, Sartaj, *Rural development : Learning from China*, Mac Millan Press, London 1978. 201 p. Biblio.

Bachrach, Peter, *Evaluating development programmes : A synthesis of recent experience*. Paris, OECD, 1977. 57 p. (OECD. Experiences in rural development. Occasional paper no. 3).

Bailey, F.G., *The peasant view of the bad life*. Brighton, University of Sussex, Institute of Development Studies, 15 p. (IDS. Communications series no. 30).

Bangladesh Academy for Rural Development, Comilla, "A new rural co-operative system for Comilla Thana", by A.K.M. Obaidullah. Rural co-operative pilot experiment, eleventh annual report, 1970-71. Comilla, Bangladesh Academy for Rural Development, 1973, 93 p.

Bates, Robert H., *Rural responses to industrialization*. New Haven, Conn., Yale University Press, 1976. x, 380 p.

Beckford, George L., *Persistent poverty : Underdevelopment in plantation economies of The Third World*. New York, Oxford University Press, 1972. 303 p. Biblio.

Bergmann, Herbert and Nasser Khademam, *The impacts of large-scale farms on development in Iran : A case study of certain aspects of the Iranian agrarian reform*. Saarbrucken, Verlag der SSIP-Schriften for the Research Centre for International Agrarian Development, Heidelberg, 1975. 204 p. Biblio.

Bhalla, G.S., *The impact of the green revolution on rural structural change. Colloquium on rural-urban relations and development planning in Asia*, 1-18 November 1977, Nagoya, Japan. Nagoya, UNCRD, 1977. 48 p.

Bhasin, Kamla, *Participatory training for development*. "Report of the Freedom from Hunger Campaign/Action for Development Regional Change Agents Programme, April-

May 1976." Bangkok, Food and Agriculture Organisation of the United Nations, 1977.

Biggs, Stephen D., et al., *Agricultural sectoral analysis*. New York, Agricultural Development Council, 1977. 29 p. (ADC. Teaching and research forum no. 11).

Blackton, John Stuart, *Local government and rural development in Sri Lanka*. Ithaca, N.Y., Cornell University, Center for International Studies, Rural Development Committee, 1974. 68 p. (Cornell University, Center for International Studies, Rural Development Committee. Rural Local Government no. 14). Biblio.

Blair, Harry W., *The elusiveness of equity: Institutional approaches to rural development in Bangladesh*, Ithaca, N.Y., Cornell University, Center for International Studies, Rural Development Committee, 1974. 127 p. Biblio.

Boserup, Ester and Christina Liljencrantz, *Integration of women in development: Why, when, how*. New York, United Nations Development Programme, 1975. 42 p. Biblio.

Boserup, Ester, *Women's role in economic development*. London, George Allen & Unwin, 1970. 283 p. Biblio.

Brown, Lester, *Seeds of change: The green revolution and development in the 1970s*. New York, Praeger for the Overseas Development Council, 1970. 205 p. Biblio.

Burki, Shadid Javed, *Agricultural growth and local government in Punjab, Pakistan*. Ithaca, N.Y., Cornell University, Center for International Studies, Rural Development Committee, 1974. 68 p. (Cornell University, Center for International Studies, Rural Development Committee. Rural Local Government no. 11). Biblio.

Cairncross, Alec (ed.), *Employment, income distribution and development strategy: Problems of the developing countries*. Essays in honour of H.W. Singer. London, Macmillan, 1976. 264 p. Biblio.

Castillo, Gelia, T., *All in a grain of rice*. Laguna, Southeast Asian Regional Center for Graduate Study and Research in Agriculture, 1975. 410 p. Biblio.

Central Treaty Organization, Ankara, *Seminar on integrated rural development*, Islamabad, 1975. (Papers). Ankara, 1975. 345 p.

Chambers, Robert, *Managing rural development : Ideas and experience from East Africa.* Uppsala, The Scandinavian Institute of African Studies, 1974. 216 p. Biblio.

Chambers, Robert, *Two frontiers in rural management : Agricultural extension and managing the exploitation of communal natural resources.* Brighton, University of Sussex, Institute of Development Studies, 1975. 12 p. (IDS Communication 113), Biblio.

Chee, Stephen, Local institutions and rural development in Malaysia. Ithaca, N.Y., Cornell University, Center for International Studies, Rural Development Committee, Rural Local Government no. 9), Biblio.

Cheema, G. Shabbir, *Rural organizations and participation in Malaysia.* Prepared for the World Conference on agrarian and Rural Reform, FAO. Rome, Italy, 1979.

Chenery, Hollis. et al., *Redistribution with growth.* London, Oxford University Press for the World Bank and the Institute of Development Studies, 1974. 304 p. Biblio.

Chitrakar, Jagadish K., *The role of panchayats in Nepal's development.* Madison, Wisc., University of Wisconsin, Center for Development, 1977. 41 p. Biblio.

Chuchart, Chaiyong, The determination and analysis of policies to support and stabilize agricultural price and incomes of the Thai farmers (with special reference to rice premium). Bangkok, Ministry of National Development, 1965. 76 p.

Connell, J. and M. Lipton, *Assessing village labour situations in developing countries.* New Delhi, Oxford University Press, 1977. 200 p.

Connell, John., *The evolution of Tanzanian rural development.* Brighton, University of Sussex, Institute of Development Studies, 1973. 21 p. (IDS Communication 110), Biblio.

Connell, John. et al., *Migration from rural areas : The evidence from village studies.* Delhi, Oxford University Press, 1976. vii, 228 p. Biblio.

Coombs, Philip Hall and M. Ahmed, *Attacking rural poverty : How informal education can help.* A research project for the World Bank prepared by the International Council for Education Development. Baltimore, Md., The Johns Hopkins University Press, 1974. 292 p.

Cornell University, Center for International Studies, Rural Development Committee, *Training and research for extended rural development in Asia.* Ithaca, New York, 1974. ix, 119 p.

Cummings, Ralph W., Minimum Information System for Agricultural Development in Low-Income Countries. New York, The Agricultural Development Council, 1977. 13 p. (ADC. Seminar report no. 14).

Dasgupta, Biplab. *Agrarian change and the new technology in India.* Geneva, United Nations Research Institute for Social Development, 1977. xxvii, 408 p.

Davidson, Joan and Gerald Wibberley, *Planning and the rural environment*, Oxford, Pergamon Press, 1977. xii, 225 p.

Day, Richard H. and Inderjit Singh, *Economic development as an adaptive process : The green revolution in the Indian Punjab*, Cambridge, Cambridge University Press, 1977. x. 326 p. Biblio.

Development Alternatives, Inc., Washington, D.C., *Strategies for small farmer development : An empirical study of rural development projects.* Washington, D.C., Development Alternatives Inc., for AID, 1975. 3 v.

Edwards, Edgar O. (ed.), *Employment in developing nations : A report on a Ford Foundation study.* New York, Columbia University Press, 1974. 428 p. Biblio.

Elliott, Charles and F. de Moster, *Patterns of poverty in the Third World : A study of social and economic stratification*, New York, Praeger for the World Council of Churches, 1975. 416 p. Biblio.

Epstein, T. Scarlett, *South India : yesterday, today and tomorrow : Mysore villages revisited*, London, Macmillan, 1973. 273 p., Biblio.

Etienne, Gilbert, *Studies in Indian agriculture : The art of the possible*, Berkeley, Calif., University of California Press, 1968. 343 p., Biblio.

Evers, Hans-Dieter (ed.) *Modernization in South-East Asia.* Singapore, Oxford University Press for the Institute of Southeast Asian Studies, 1973. 249 p. Biblio.

Farmer, B.H. (ed.), *Green revolution ? Technology and change in rice-growing areas of Tamil Nadu and Sri Lanka.* London, Macmillan, 1977. xvi, 429 p.

Fisk, E.K., *Studies in the rural economy of South-East Asia.* Singapore, Eastern Universities Press, 1964. 108 p.

Flores, Xavier, *Agricultural organisations and economic and social development in rural areas*, Geneva, ILO, 1971. 586 p. Biblio.

Food and Agriculture Organization of the United Nations, *FAO/SIDA Expert consultation on policies and institutions for integrated rural development*, Nairobi, 1976, Rome, FAO, 1977, 2 v.

Food and Agriculture Organization of the United Nations, *Joint report on the expert consultations on policies and institutions for integrated rural development*, Colombo, Sri Lanka, 20-30 October 1975 and Jakarta, Indonesia, 2-12 December 1975. Rome, FAO, 1977.

Food and Agriculture Organization of the United Nations, *Planning and evaluation of applied nutrition programmes*, by Michael C. Latham. Rome, FAO, 1972. 125 p. (FAO Nutritional studies, 26).

Food and Agriculture Organization of the United Nations, Bangkok, *Report of the expert group meeting on agrarian reform and rural development*, Bangkok, FAO, 1978. 34 p., append.

Food and Agriculture Organization of the United Nations, Rome, *Training for agriculture and rural development*, 1977. Rome, FAO, 1977. 122 p. (FAO Economic and Social Development series no. 7).

Food and Agriculture Organization of the United Nations, Regional Office For Asia and the Far East, *Learning from China : A report on agriculture and the Chinese people's communes*, 9 September—5 October 1975. Bangkok, 1977. 112 p.

Friedrich-Ebert-Stiftung, (ed.), *Effective anti-poverty strategies*, Regional experts, workshop organized by Friedrich-Ebert-Stiftung in co-operation with ECAFE, Bangkok, 12-21 December 1973. Bangkok, FES, 1973. 332 p. (FES Workshop series report no. 4).

Friedrich-Ebert-Stiftung, (ed.), *Organization of Peasants in Asia*, Regional experts, workshop organized by Friedrich-Ebert-Stiftung, U.N. FAO and ILO, 28 October—2 November 1974, Bangkok. Bangkok, FES, 1974. 316 p.

Gebregziabher, Betru, *Integrated development in rural Ethiopia : An evaluative study of the Chilalo Agricultural Development Unit*, Bloomington, Ind., Indiana University, International Development Research Center, 1975. xv, 78 p. Biblio.

Ghai, D.P., et al., *The Basic-needs approach to development : Some issues regarding concepts and methodology*, Geneva ILO, 1977. 113 p.

Goulet, Denis, *The cruel choice : A new concept in the theory of development*. New York, Atheneum, 1973. 360 p.

Griffin, Keith B., *Land concentration and rural poverty*, New York, Holmes and Meier Publishers, 1976. viii, 303 p.

Griffin, Keith, *The political economy of agrarian change : An essay on the green revolution*. Cambridge, Mass., Harvard University Press, 1974. 264 p., Biblio.

Hansen, Gary E., *Rural local government and agricultural development in Java, Indonesia*. Ithaca, N.Y., Cornell University, Center for International Studies, Rural Development Committee, 1974. 78 p. (Cornell University, Center for International Studies, Rural Development Committee. Rural Local Government no. 7). Biblio.

Harkin, Duncan A., *Philippine agrarian reform in the perspective of three years of martial law*. Madison, Wisc., University of Wisconsin-Madison, Land Tenure Center, 1976. 30 p. (A research paper no. 68).

Haq, Mahbub Al, *The poverty curtain : Choices for the Third World*. Columbia University Press, 1976.

Haswell, Margaret R., *Tropical farming economics*, London. Longman, 1973. 174 p. Biblio.

Herring, Ronald, and M. Ghaffer Chaudhary, *The 1972 land reforms in Pakistan and their economic implications : A preliminary analysis*. Madison, Wisc., University of Wisconsin, Land Tenure Center, 1975. 245-279 p. (LTC Reprint no. 126).

Hewes, Laurence, *Rural development : World frontiers*. Ames, Iowa, The Iowa State University Press, 1974. 186 p.

Hirshima, Shigemochi, *The structure of disparity in developing agriculture : A case study of the Pakistan Punjab*. Tokyo, IDE, 1978. x, 138 p. (IDE. Occasional papers no. 16). Biblio.

Hopper, W. David, *Rural development political change*, Ottawa, International Development Research Center, 1972. 15 p.

Huizer, Gerrit, *Agrarian unrest and peasant organizations in the Philippines*. The Hague, Institute of Social Studies, 1972. 26 p. (ISS occasional papers no. 17).

Hunter, Guy, *The administration of agricultural development : Lessons from India*. London, Oxford University Press, 1970. 160 p.

Hunter, Guy, *The implementation of agricultural development policies : Organization, management and institutions*, Summary and comment on the Second International Seminar on Change in Agriculture, at the University of Reading, 9-19 September 1974. London, Overseas Development Institute, 1974. 39 p.

Hunter, Guy, *Modernizing peasant societies : A comparative study in Asia and Africa*. London, Oxford University Press for the Institute of Race Relations, 1969. 324 p.

Hunter, Guy (ed.), *Policy and practice in rural development*. London, Croom Helm for Overseas Development Institute, 1976. 520 p. Biblio.

Hunter, Guy and Reve Dermont, *Rural development in manpower aspects of educational planning : Problems for the future*, Paris, UNESCO/IIEP, 1968.

Hunter, Guy, *Serving the small farmer : Policy choices in Indian agricultural development*. London, Croom Helm, 1974. 250 p.

Huq, M. Ameerul (ed.), *Exploitation and the rural poor*. Comilla, Bangladesh Academy for Rural Development, 1976.

Iglesias, Gabriel U., *The national rice self-sufficiency program of the Philippines : A case study in the implementation of a development program*. Quezon City, University of the Philippines, Office of the Vice-President for Academic Affairs, Office of Research Co-ordination, NSDB-UP Integrated Research Program, 1976. 85 p. Biblio.

Irayatullah (ed), *Approaches to rural development : Some Asian experiences*. Asian and Pacific Development Administration Centre, Kuala Lumpur, Malaysia, 1979. 479 p.

Imboden, Nicolas, *Planning and design of rural drinking water projects*, Paris, OECD, 1977. viii. 51 p. (OECD. Experiences in Rural Development. Occasional paper no. 2)

Imboden, N., *Management approach to project appraisal and evaluation with special reference to non-directly productive enterprises*. Development Centre of OECD, Paris, 1978.

India, Planning Commission, Report of the working group on block level planning. New Delhi, 1978.
Asoka Mehta Committee Report

Indian Society of Agricultural Economics, Bombay, *Problems of farm mechanization*. Bombay, 1972. 179 p. (Seminar series 9).

Indian Society of Agricultural Economics, Seminar 13. Bangalore, 1974. *Role of irrigation in the development of India's agriculture*. Bombay, 1976. xx, 201 p.

Ingle, Marcus D., *Local governance and rural development in Thailand*, Ithaca, N.Y., Cornell University, Center for International Studies, Rural Development Committee, 1974, 96 p (Cornell University, Center for International Studies, Rural Development Committee. Rural Local Government no. 16). Biblio.

Institute of Applied Manpower Research, New Delhi, *Manpower development in rural India : A case study*, New Delhi, Allied Publishers Private Ltd., 1977. 107 p.

Institute of Human Ecology, Laguna, *An annotated bibliography on rural development*. Laguna, Institute of Human Ecology, 1975. 62 p.

International Bank for Reconstruction and Development, Washington, D.C. *Rural enterprise and nonfarm employment*. Washington, D.C., IBRD, 1978. 87 p.

International Bank for Reconstruction and Development, Washington, D.C. "Basic needs: An issues paper", prepared by P. Streeten and S J. Burki. Washington, D.C., 1977. 21 p.

International Cooperative Alliance, "Report of the national seminar on integrated approach to agricultural cooperative development in Nepal", Kathmandu, Nepal, 5—14 December 1976. New Delhi, ICA and Kathmandu, Department of Cooperatives, 1977. 22 p.

International Development Research Center : "Integrated approach to local rural development". Report of an interdisciplinary seminar, Makati, Philippines, 31 March—3 April 1975. Edited by Marilyn Campbell. Ottawa, IDRC, 1975. 52 p.

International Labour Office, "National seminar on landless labour and job opportunities in rural areas". Dacca, 8—10 November 1977, Report. Dacca, ILO, 1977. 197 p.

International Labour Office, Geneva *Employment, growth and basic needs : A one-world problem.* Geneva, ILO, 1976. 177 p.

International Labour Office, Geneva, *Meeting basic needs : Strategies for eradicating mass poverty and unemployment.* Geneva, ILO, 1977. 60 p.

International Labour Office, Geneva *Poverty and landlessness in rural Asia.* Geneva, ILO, 1977. 288 p. Biblio.

International Labour Office, Geneva *Structure and functions of rural workers' organizations : A workers' education manual.* Geneva, ILO, 1978. ix, 159 p.

Islam, Nurul (ed.), *Agricultural policy in developing countries : Proceedings of a conference held by the International Economic Association at Bad Godesberg, West Germany.* New York, John Wiley, 1974. 565 p.

Jacobs, Norman, *Modernization without development : Thailand as an Asian case study.* New York, Praeger, 1971. 420 p., Biblio.

Jacoby, Erich H. and C.F. Jacoby, *Man and land : The essential revolution.* New York, Alfred A. Knopf, 1971. 400 p. Biblio.

Jain, Netra Pal. *Rural reconstruction in India and China : A comparative study,* New Delhi, Writers and Publishers Corporation, 1970. xx, 370 p.

Johl, S.S. and Mohinder S. Mudahar, *The dynamics of institutional change and rural development in Punjab, India,* Ithaca, N.Y. Cornell University, Center for International Studies, Rural Development Committee, 1974. xi, 171 p. (Cornell University, Center for International Studies, Rural Development Committee. Rural Local Government no. 5). Biblio.

Joshi, P.C., *Land reforms in India : Trends and perspectives.* Bombay, Allied Publishers for the Institute of Economic

Growth, Delhi, 1975. 181 p. (Institute of Economic Growth. Studies in Economic Growth no. 19), Biblio.

Kamarck. Andrew M., *The tropics and economic development : A provocative inquiry into the poverty of nations*. London, Johns Hopkins University Press, 1976, xiv, 113 p.

Khan, Akhter Hameed, *Reflections on the Comilla rural development projects*, Washington, D.C., American Council of Education, 1974. 46 p. (American Council on Education, Overseas Liaison Committee, paper 3.)

Khan, Mahmood Hasan, *The economies of the green revolution in Pakistan*. New York, Praeger, 1975. xvi, 227 p.

Klatzmann, Joseph. et al. (ed.), *The role of group action in the industrialization of rural areas*, New York, Praeger Publishers, 1971. 599 p.

Kocher, James E., *Rural development, income distribution and fertility decline*, New York, The Population Council, 1973. 105 p. Biblio.

Korea, Ministry of Home Affairs, *Saemaul Undong*. Seoul, 1978, 84 p.

Korea, Office of Rural Development, *Rural development program in Korea*, Suweon, 1977. 1 v.

Krishna, Raj, *Rural unemployment : A survey of concepts and estimates for India*. Washington, D.C., World Bank, 1976. 67 p. World Bank staff working paper no. 234).

Kulp, Earl M., *Rural development planning : Systems analysis and working method*, New York, Praeger, 1970. xxi, 664 p. Biblio.

Kuitenbrouwer, J.B.W., *Towards Self-Reliant Integrated Development*. The Hague, Institute of Social Studies, 1975. 44 p, (ISS Occasional papers no. 55).

Kuo, Leslie Tse-chiu, *Agriculture in the People's Republic of China*, New York, Praeger, 1976. xx, 288 p. Biblio.

Ladejinsky, Wolf, *The green revolution in Bihar—the Kosi area : A field trip, The green revolution in Punjab : A field trip*. New York, Agricultural Development Council, 1976. 20 p, (ADC Report 28).

Leagans, J. Paul and Charles P. Loomis (ed.), *Behavioural change in agricultuve : concepts and strategies for influencing transition*. Ithaca, N.Y., Cornell University Press, 1971, xii, 506 p.

Lehmann, David (ed.), *Agrarian reform, and agrarian reformism: Studies of Peru, Chile, China and India*, London, Faber and Faber, 1974. 320 p. Biblio.

Lele, Uwa, *The design of rural development : Lessons from Africa*, Baltimore, Md., Johns Hopkins University Press, 1975. xiii, 246 p.

Lesesma, Antonio J., *Land reform programs in East and Southeast Asia : A comparative approach*. Madison, University of Wisconsin-Madison, Land Tenure Center, 1976. 60 p. (A research paper no. 69).

Lipton, Michael, *Why people stay poor : A study of urban bias in world development*. London, Temple Smith, 1977. 467 p. Biblio.

London, Paul A., *Merchants as promoters of rural development : An Indian case study*, New York, Praeger, 1975. xii, 159 p. Biblio.

Lynch, Frank S.J., *Field data collection in developing countries : Experiences in Asia*. A/D/C seminar report. 1—20 June 1976. New York, The Agricultural Development Council Inc.

Malassis, Louis, *The rural world : Education and development*, London, Croom Helm, 1976. 128 p.

Mann, Amy G. and J. Miracle (ed.), *Rural development : The interplay of analysis and action*, Bloomington, Ind., Indiana University, IDRC, 1975. 67 p.

Mannan, M.A., *Rural leadership and its emerging pattern in Bangladesh : With special reference to Comilla Kotwali Thana*,Comilla, Bangladesh Academy for Rural Development, 1972. 30 p.

Marga Institute, Colombo, *The co-operative system and rural credit in Sri Lanka*, Colombo, Marga Institute, 1974. 174 p. (Marga research studies 3). Biblio.

Mosher, A.T., *To create a modern agriculture : Organization and planning*, New York, Agricultural Development Council, 1971. 162 p. Biblio.

Mosher, A.T., *Thinking about rural development*, New York, Agricultural Development Council, 1976. 250 p.

Mosher, A.T., *Projects of integrated rural development*, New York, Agricultural Development Council, 1972, 8 p.

Morss, Eliott R., *Implementing the US aid—McNamara mandate*, *what big foreign donors can do about getting the benefits of rural development to the small farmer*, Ithaca, N.Y., Cornell University Press, 1974. mimeo.

Morss, Elliott R., et al., *Strategies for small farmer development*, Study for IAD, Boulder, Colo., Westview Press, 1976, 2 v.

Myrdal, Gunnar, *Asian drama : An inquiry into the poverty of nations*, New York, The Twentieth Century Fund, 1968, ev.

Myrdal, Gunnar, *The challenge of world antipoverty program in outline*, London, Allen Lane the Penguin Press, 1970, 518 p.

National Academy of Sciences, Washington, D.C. *Energy for rural development : Renewable resources and alternative technologies for developing countries*, Washington, D.C., NAS, 1976, 306 p. Biblio.

Nerfin, Marc (ed.), *Another development : Approaches and strategies*, Uppsala, The Dag Hammarskjold Foundation, 1977, 265 p.

Ness, G.D., *Bureaucracy and rural development in Malaysia.* Berkeley, Calif., University of California Press, 1967, 257 p. Biblio.

Nicholson, Norman K. and Dilawar Ali Khan, *Basic democracies and rural development in Pakistan*, Ithaca, N.R., Cornell University, Center for International Studies, Rural Development Committee, 1974. 95 p. (Cornell University. Center for International Studies. Rural Development Committee. Rural Local Government no. 10). Biblio.

North Central Regional Center for Rural Development, *Rural development : Research priorities*, edited by L.R. Whiting, Ames, Iowa, The Iowa State University Press, 1973, 113 p.

Nulty, Leslie, *The green revolution in West Pakistan : Implications of technological change*, New York, Praeger, 1972, 150 p. Biblio.

Obaidullah Khan, A.Z.M., *Poverty-oriented rural development and the U.N. system : A turning point*, with collaboration of J. Tomas Hexner. A working document for the ACC.

Ollawa, Patrick, *Rural development policies and performance in Zambia : A critical inventory*, The Hague, Institute of Social Studies, 1977. 43 p. (ISS Occasional papers no. 59).

Onchan, Tongroj, *Agricultural credit policy in Asian countries*, Bangkok, 1977, 38 p. Mimeo. (Kasetsart University, Faculty of Economics and Business Administration, staff paper, 20).

Onchan, Tongroj, *Agricultural credit problems in Thailand*, New York, Agricultural Development Council, 1971, 104 p, (ADC. National seminar report no. 1).

Onchan, Tongroj and Leonardo Paulino, *Rural poverty, income distribution and employment in developing Asian countries*, Bangkok, Kasetsart University, 1977, 95 p. Mimeo.

Owen, Wyn F., *Two rural sectors : Their characteristics and roles in the development process*, Madison, Wisc., University of Wisconsin, Land Tenure Center, 1974. 403-437 p. (LTC Reprint 118).

Paige, Jeffery M., *Agrarian revolution : Social movements and export agriculture in the underdeveloped world*, New York, The Free Press, 1975. 435 p. Biblio.

Pakistan, Ministry of Food and Agriculture, *Agricultural enquiry committee : Report of the working group on agricultural credit, cooperatives and integrated rural development programme*, Islamabad, 1975, 69 p.

Palmer, Ingrid, *The new rice in the Philippines*, Geneva, UNRISD, 1975, 200 p. Biblio.

Parbery, David B., *The comprehensive rural development package program (COMPAC) for Thailand*, Honolulu, Hawaii, East-west Center, 1971. 81 p.

Pearson, Lester B., *Partners in development : Report of the Commission on International Development*, New York, Praeger, 1969. 399 p.

Polemen, Thomas T. and D.K. Freebairn (ed), *Food, population, and employment : The impact of the green revolution*, New York, Praeger for the Cornell University Program on Science, Technology and Society, 1973. 272 p.

Qadeer, Mohammed A., *An evaluation of the integrated rural development programme*, Islamabad, Pakistan Institute of Development Economics, 1977. 95 p. Qad.

Rao, C.H. Hanumantha, *Technological change and distribution of gains in Indian agriculture*, Delhi, Institute of Economic Growth. 1975, xiv, 249 p.

Rao, V.K.R.V., *Growth with justice in Asian agriculture : An exercise in policy formulation*, Geneva, United Nations Research Institute for Social Development, 1974, 96 p.

Raper, Arthur F., *Rural development in action : The comprehensive experiment at Comilla, East Pakistan*. Ithaca, N.Y., Cornell University Press, 1970. 351 p., Biblio.

Rawski, Evelyn Sakakida, *Agricultural change and the peasant economy of South China*, Cambridge Mass., Harvard University Press, 1972. 282 p. Biblio.

Reynolds, Lloyd G. (ed.), *Agriculture in development theory*, New Haven, Conn., Yale University Press, 1975. x, 510 p., Biblio.

Robinson, Edward Austin George and Keith Griffin (ed.), *The economic development of Bangladesh within a socialist framework*, Proceedings of a conference held by the International Economic Association at Dacca, 1973, New York, John Wiley. 1974, 330 p.

Rogers, Everett M., *Modernization among peasants : The impact of communication*, New York, Holt, Rinehart and Winston, 1969. 429 p.

Rondinelli, Dennis A. and K. Ruddle, *Urban functions in rural development : An analysis of integrated spatial development policy*, Washington, D.C., U.S. Dept. of State, AID, Technical Assistance Bureau, Office of Urban Development, 1976., 305 p. Biblio.

Rosen, George, *Peasant society in a changing economy : Comparative development in Southeast Asia and India*, Urbana, ILL., University of Illinois Press, 1975, 256 p., Biblio.

Roumasset, James A., *Rice and risk : Decision making among low-income farmers*, Amsterdam, North-Holland, 1976. xxi, 251 p. Biblio.

Rubin, Herbert J., *The dynamics of development in rural Thailand*, DeKalb, Northern Illinois University, Center for Southeast Asian Studies, 1974, x, 156 p., Biblio.

Saihoo, Pathya, *A report on leadership in the southern provinces*, Bangkok, Community Development Division and the Faculty of Political Science, Chulalongkorn University, 1965. 678 p.

Schumacher, E.F., *Small is beautiful : Economics as if people mattered*, New York, Harper and Row, 1973, 305 p. Biblio.

Schwefel, Detlef, *Basic needs planning and evaluation*, German

Development Institute, Berlin 1978, 358 p. (Occasional Paper no. 50.) Biblio.

Schwenk, Richard L., *The potential for rural development in the new Seventh Division of Sarawak : A preliminary background report*, Singapore, Institute of Southeast Asian Studies, 1973, 38 p. (ISEAS. Occasional paper no. 17). Biblio.

Scott, James C. and Benedict J. Kerkvliet, *How traditional rural patrons lose legitimacy*, Madison, Wisc., University of Wisconsin, Land Tenure Center, 1975. 501-540 p. (LTC Reprint no. 125).

Scott, Wolf, *Measurement and analysis of progress at local level : An overview*, United Nations Research Institute for Social Development, Geneva, 1978.

Sevele, F.V., *Rural Employment and Rural Development in the Pacific Islands*, South Pacific Commission, Noumea, New Caledonia Feb. 1979. Occasional paper no. 13.

Sen, Amartya, *Employment, technology and development : A study prepared for the International Labour Office within the framework of the World Employment Programme*, Oxford, Clarendon Press, 1975, 193 p.

Sen, Sudhir, *Reaping the green revolution : Food and jobs for all*, Maryknoll, N.Y., Orbis Books, 1975. 397 p., Biblio.

Sen, Sudhir, *A richer harvest : New horizons for developing countries*, Marknoll, N.Y., Orbis Books, 1974, 573 p. Biblio.

Shand, R T., *Technical change in Asian agriculture*, Canberra, Australian National University Press, 1973, xxi, 319 p. Biblio.

Simpas, Santiago, and others, *Local government and rural development in the Philippines*, Ithaca, N.Y., Cornell University, Center for International Studies, Rural Development Committee, 1974. 109 p. (Cornell University, Center for International Studies, Rural Development Committee. Rural Local Government no. 12). Biblio.

Singh, Gajendra (ed.) and J.H. de Geode (ed.), *Rural development technology : An integrated approach*, Proceedings of the International Conference held at Bangkok on 21-24 June 1977. Bangkok, Asian Institute of Technology, 1977, 847 p.

Sinha, Radha, *Food and poverty : The political economy of confrontation*, New York, Holmes and Meier, 1976. 196 p.

Snow, Edgar, *The long revolution*, New York, Random House, 1971, 269 p.

Southeast Asia Development Advisory Group, Asia Society, *New York, Agricultural revolution in Southeast Asia*, New York, 1970, 2 v.

Southeast Asia Development Advisory Group New York, *Models of rural development administration : The JCCR experience in Taiwan*, by R.L. Hough. New York, the Asia Society, 1968, 24 p.

Southeast Asia Development Advisory Group, New York, *Models of rural development administration : The Malaysian Ministry for rural development*, by G.D. Ness. New York, The Asia Society, 1967. 7 p.

Srinivas, M.N., *Science, technology and rural development in India*, Bombay Orient Longman, 1977. 15 p.

Stavis, Benedict, *Making green revolution : The politics of agricultural development in China*, Ithaca, N.Y., Cornell University, Center for International Studies, Rural Development Committee, 1974, ix, 287 p. Biblio.

Stavis, Benedict, *People's communes and rural development in China*, Ithaca, N.Y., Cornell University, Center for International Studies, Rural Development Committee, 1974, vi, 173. Biblio.

Stevens, Robert D. et al. (eds.) *Rural development in Bangladesh and Pakistan*, Honolulu, Hawaii, The University Press of Hawaii, 1976, xiii, 399 p. Biblio.

Subramaniam, C., *A strategy for rural development*, Bangkok, U.N. Asian and Pacific Development Institute, 1975, 15 p.

Sunthornphesuth, Suthep, *An intensive study of a village on local leadership, Amnatcharoen district, Ubol province, research report*, Bangkok, Dept. of Community Development and the Social Science Dept. of the Faculty of Political Science, Chulalongkorn University, 1968, 94 p.

Tai, Hung-Chao, *Land reform and politics : A comparative analysis*, Berkeley, Calif., University of California Press, 1974. xxi, 565 p. Biblio.

Thailand, Ministry of Interior, Community Development Department, Research and Evaluation Division *Report on the evaluation of rural development volunteers program*, Bangkok, 1971, 96 p.

Thailand, Ministry of Interior. Department of Community Development. Research and Evaluation Division. "A research report on the study of social and economic conditions of the people in a development area," Tung Luknog Commune, Kampaengsan District, Nakorn Patom Province, 1968, 81 p.

Thailand, Ministry of Interior. Office of Accelerated Rural Development. Planning and Project Division, Fact sheets on Thailand : rural development, Bangkok. Public Relations Department, 1978. 6 p.

Thomas, John Woodward and R.M. Hook, *Creating rural employment : A manual for organizing rural works programs*, Washington, D.C., AID, 1977, 127 p., Biblio.

United Nations, Asian Development Institute, Bangkok "An approach to evolving guidelines for rural development". A report on the expert group meeting, December 9 - 13, 1970. Bangkok, 1975. 67 p.

United Nations, Asian Development Institute, Bangkok, *Towards a theory of rural development* by Wahidul Haque, Niranjan Mehta, Anisur Rahman and Ponna Wignaraja. Bangkok, 1975, 102 p.

United Nations, Department of Economic and Social Affairs, New York, *Evaluation of the binational rural housing demonstration project in Columbia and Ecuador*, New York. UN, 1977. 60 p.

United Nations, Department of Economic and Social Affairs, New York, *Popular participation in decision making for development*, New York, UN, 1975, 65 p.

United Nations, Department of Economic and Social Affairs, New York, *Popular participation in development : Emerging trends in community development*, New York, UN, 1971, 273 p.

United Nations, Department of Economic and Social Affairs, New York, *Poverty, unemployment and development policy : A case study of selected issues with reference to Kerala*, New York, UN, 1975, 235 p.

United Nations, Department of Economic and Social Affairs, New York, *Rural housing : A review of world conditions*, New York, UN, 1969, 186 p.

United Nations. Department of Economic and Social Affairs,

New York, *The significance of rural housing in integrated rural development*, New York, UN, 1978, 117 p. Biblio.

United Nations, Economic Commission for Africa, *Integrated approach to rural development in Africa*. New York, UN, 1971, 115 p.

United Nations, Economic and Social Commission for Asia and the Pacific, Bangkok, *Premises and implications of a unified approach to development analysis and planning*, by Joseph B.W. Kuitenbrouwer. Bangkok, ESCAP, 1976, 42 p. Biblio.

United Nations, Food and Agriculture Organization, Rome. *Report on the FAO/SIDA/DSE inter-regional symposium on integrated rural development*, held in Berlin, 19 - 23 September 1977. Rome, FAO, 1978. 121 p.

United Nations, Research Institute for Social Development, Geneva, *Rural institutions as agents of planned change*, edited by Orlando Fals Borda and C. Inayatullah. Nos. 1, 4-8. Geneva, 1969-1975.

United States, Agency for International Development, Washington, D.C., *Small farmer credit in East Asia : The BIMAS Program in Indonesia, selected papers*, Washington, D.C., 1973, iv.

United States, Agency for International Development, Washington, D.C. *small farmer credit in South Asia : Small farm credit in Bangladesh*, by M. Solaiman and Azizul Huq. Washington, D.C., AID, 1973., i.v.

Uphoff, Norman T. and Milton J. Esman, *Local organization for rural development : Analysis of Asian experience*, by Ithaca, N.Y., Cornell University, Center for International Studies, Rural Development Committee, 1974. xxiv, 154 p. (Cornell University, Rural Local Government series no. 19).

Van Roy, Edward, *Economic systems of Northern Thailand : Structure and change*. Ithaca, N.Y., Cornell University Press, 1971. 289 p., Biblio.

Veit, Lawrence A., *India's second revolution : The dimensions of development*, New York, McGraw-Hill, 1976, xii, 403 p. Biblio.

Walinsky, Louis J. (ed.), *Agrarian reform as unfinished business*. New York, Oxford University Press, 1977, xi, 603 p.

Weiss, Carol H. (ed.), *Evaluating action programmes, readings in social action and evaluation.* Boston, Mass., Allyn and Bacon, 1972, 338 p.

Weiss, Carol H., *Evaluation research : Methods for program effectiveness.* Englewood Cliffs, N.J., Prentice-Hall, 1972. 160 p.

Weitz, Raanan, *From peasant to farmer : A revolutionary strategy for development.* New York, Columbia University Press, 1971., 292 p.

Weitz, Raanan (ed.), *Rural development in a changing world.* Cambridge, Mass., MIT Press, 1971, 587 p.

Whyte Williams F. and L.K. Williams, *Toward an integrated theory of development : Economic and non-economic variables in rural development*, Ithaca, N.Y., Cornell University, 1968. 89 p.

Wichaidit, T., *Provincial administration in Thailand : Its development and present problems,* Ann Arbor, Mich., University Microfilms, 1973, 289 p. Biblio.

Winkelman, Don, *The traditional farmer : Maximization and mechanization*, Paris, OECD, 1972, 94 p. Biblio.

Wong, John, *Chinese land reform in retrospect*, Madison, Wisc., University of Wisconsin, Land Tenure Centre, 1974, 32 p. Biblio. (LTC Reprimt no. 113).

Wong, John and Edward D. Reed, *The experience and potential for group farming in Asia,* New York, Agricultural Development Council, 1978, 17 p.

Wong, John, *Land reform in the People's Republic of China : Institutional transformation in agriculture*, New York, Preager, 1973, 317 p. Biblio.

World Bank, Washington, D.C. *The assault on world poverty : problems of rural development, education and health*: Baltimore, Md :, The Johns Hopkins University Press for the World Bank, 1975, xi, 425 p.

World Bank, Washington, D.C., *The economic analysis of rural road projects,* Washington, D.C., IBRD, 1976, 1 v. (World Bank staff working paper no. 241).

World Bank, Washington, D.C., *Rural development : Sector policy paper*, Washington, D.C., 1975, 89 p.

World Bank, Washington, D.C., *Rural Electrification.* Washington, D.C., World Bank, 1975, 79 p.

World Bank, Washington, D.C., *Village water supply : A world Bank paper*. Washington, D.C., World Bank, 1976, 96 p.

Wurfel David, *Philippine agrarian policy today : Implementation and political impact*. Singapore, Institute of Southeast Asian Studies, 1977, 41 p., (Institute of Southeast Asian Studies, occasional paper no. 46), Biblio.

Yudelman, Montague, et al., *Technological change in agriculture and employment in developing countries*, Paris, OECD, 1971, 204 p.

Zaman, M.A., *Land reform in Bangladesh to 1970*, Madison, Wisc., University of Wisconsin, Madison, Land Tenure Center, 1976. 51 p., (A research paper no.66).

ARTICLES

Ahluwalia, Montek S., "Rural poverty and agricultural performance in India, *The Journal of Development Studies*, April 1978, 14 (3) : 298-323.

Bardhan, P.K., "On measuring rural unemployment," *The Journal of Development Studies*, April 1978, 14(3) : 342-352.

Bell, Clive, "Alternative theories of sharecropping : Some tests using evidence from Northeast India," *The Journal of Development Studies*, July 1977, 13(4) : 317-346.

Blair, Harry W., "Rural development, class structure and bureaucracy in Bangladesh, *World Development*, January 1978, 6(1) : 65-82.

Brown, David W., "Rural development from a decision-making perspective," *International Development Review*, 17(2) : 12-16.

Clayton, Eric S., "Programming rural employment opportunities in Kenya," *International Labour Review*, August-September 1975, 112 (2-3) : 149-161

Das Gupta, Malabika, "Distribution of benefits from growth in rural India," *margin*. July 1978, 10(4) : 34-43

Davies, Morton R., "Choices in agricultural development programmes," *International Review of Administrative Sciences*, 1977, 43(3) : 197-204.

Davis-Friedmann, Deborah, "Welfare practices in rural China. *World Development*, May 1978, 6(5) : 609-619.

Dawson, A., "Suggestions for an approach to rural development by foreign aid programmes," *International Labour Review*, July-August 1978, 117(4) : 391-404.

Dore, R.P., "Learning from China : Reflections on transferability." *World Development*, May 1978, 6(5) : 715-717.

Galtung, Johann, "What is a strategy" *International foundation of development alternatives*. Dossier 2, Nov. 1978, pp. 13-22.

Gersovitz, Mark, "Land reform : Some Theoretical Considerations." *The Journal of Development Studies*, October 1976, 13(1) :79-91.

Gray, Jack, "Mao and the Chinese Rural Economy," *World Development*, May 1978, 6(5) : 567-581.

Green, Reginald Herbold, Transferability, exoticism and other forms of dogmatic revisionism." *World Development*, May 1978, 6 (5) : 709-713.

Green, Reginald Herbold, "Basic human needs : A strategic conceptualization toward another development," International Foundation of Development Alternatives Dossier 2, Nov, 1978, pp. 3-12

Griffin, Keith, "Increasing poverty and changing ideas about development strategies," *Development and Change*, October 1977, 8(4) : 491-508.

Gunatilake, Godfrey, "Sri Lanka : A national dialogue on development" *International Foundation for Development Alternatives* Dossier 14, Dec. 1979, pp. 1-14

Gurley, John G., "Rural development in China 1949-72, and the lessons to be learned from it", *World Development*, July-August 1975, 3(7 & 8) : 455-471.

Huang, Yukon, "Backward-bending supply curves and behaviour of subsistence farmers," *The Journal of Development Studies*, April 1976, 12(3) : 191-11.

Inayatullah, "An analysis of the emergence of a rural development innovation in Comilla, Bangladesh, *The Journal of Developing Arcas*, October 1976, 11(1) ; 79-90.

Johnston, Bruce F. and Anthony Meyer. "Nutrition, health, and population strategies for rural development," *Economic Development and Culture Change*, October 1977, 26(1) : 1-23.

Junankar, P.N., "Land tenure and Indian agricultural policy".

The Journal of Development Studies. October 1976, 13(1) ; 42-60.

Koo, Anthony Y.C., "An economic justification for land reformation," *Economic Development and Cultural Change*, April 1977, 25(3) : 523-538.

Lele, Uma., Designing rural development programs : Lessons from past experience in Africa, *Economic Development and Cultural Change*, January 1976, 24(2) : 287-308.

Lipton, Michael, "Agricultural finance and rural credit in poor countries," *World Development*, July 1976, 4(7) : 543-553.

Malhotra, Ram C., "Development Problem of Small Island Economies with Special Reference to Rural Development", APDI Quarterly Newsletter Vol. XIII no. 3 July 1978.

Mahmood, Moazam, "The pattern of adoption of green revolution technology and its effect on land holdings in the Punjab, *Pakistan Economic and Social Review*, spring-Summer 1977, 15 (1-2) : 34-68.

Mathur, Hari Mohan, "Training administrators for rural development : Recent experience in Rajasthan, India, *International Review of Administrative Sciences*, 1977, 43(4) : 369-372.

Maxwell, Neville, "The fourth mobilization : A new phase of the Tachai movement," *Asian Survey*, April 1978, 6(4) : 499-518.

Maxwell, Neville, "Learning from Tachai," *World Development*, July-August 1975, 3(7 & 8) ; 473-495.

Moulik, T.K, "Strategies of implementation of rural development programmes in India", *International Foundation of Development Studies*, Dossier 5, March 1979, pp. 1-16.

Mukherji, I.N., "Agrarian reforms in Bangladesh." *Asian Survey*, May 1976, 16 (5) : 452-464.

Nyerere, Julius, "On rural development", *International Foundation for Development alternatives*, Dossier 11, September 1979, pp 1-15,

Overholt, William H., "Land reform in the Philippines," *Asian Survey*, May 1976, 16(5) : 427-451.

Paine, Suzanne, "Some reflections on the presence of 'rural' or of 'urban bias' in China's development policies 1949-1976." *World Development*, May 1978, 6(5) : 561-566.

Perkins, Dwight, "Meeting basic needs in the People's Republic of China," *World Development*, May 1978, 6(5) : 561-566.

Rempel, Henry, and Richard A. Lobdell, "The role of urban-to-rural remittances in rural development," *The Journal of Development Studies*, April 1978, 14(3) : 324-341.

Riskin, Carl, "China's rural Industries : Self-reliant systems or independent kingdoms ?" *The China Quarterly*, March 1978, 73 : 77-98.

Riskin, Carl, "Political conflict and rural industrialization in China," *World Development*, May 1978, 6(5) : 681-692.

Rodgers, G., "A conceptualisation of poverty in rural India." *World Development*, April 1976, 4(4) : 261-276.

Rondinelli, Dennis A. and Kenneth Ruddle, "Local organization for integrated rural development : Implementing equity policy in developing countries," *International Review of Administrative Sciences*, 1797, 43(1) : 20-30.

Roth, David F., "Dimensions of policy change : Towards an explanation of rural change policies in Thailand." *Asian Survey*, November 1976, 16(11) : 1043-1063.

Roy, Shyamal, and Melvin Blase, "Farm tractorisation, productivity and labour employment : A case study of Indian Punjab," *The Journal of Development Studies*, January 1978, 14(2) : 193-209.

Ruddle, Kenneth, "Change as perceived as man-made hazard in a rural development environment," *Development and Change*, July 1976, 7(3) : 311-330.

Ruttan, Vernon W., "Integrated rural development programs : A skeptical perspective," *International Development Review*, 1975, 17(4) : 9-16.

Schluter, Michael G.G., "Some management objectives of the peasant farmer : An analysis of risk aversion in the choice of cropping pattern, Surat District, India," *The Journal of Development Studies*, April 1976, 12(3) ; 248-61.

Seers, Dudley "The Meaning of Development" in Norman T. Uphoff and Warren f. Ilchman (ed.) *The Political Economy of Development*, Berkeley University of California Press, 1972.

Sigurdson, Jon, "Rural industrialization : A comparison of development planning in China and Japan," *World Development*, May 1978, 6(5) : 667-680.

Sigurdson, Jon, "Rural industrialization in China : Approaches and results," *World Development*, July-August 1975, 3(4) : 527-538.

Smith, Douglas V., "New seeds and income distribution in Bangladesh," *The Journal of Development Studies*, January 1975, 11(2) : 162-186.

Stavis, Benedict, "Agricultural research and extension services in China," *World Development*, May 1978, 6(5) : 631-645.

Stavis, Benedict, China's rural local institutions in comparative perspective," *Asian Survey*, April 1976, 16(4) : 381-396.

Stryker, J. Dirck, "Population density, agricultural technique and land utilization in a village economy," *The American Economic Review*, June 1976, 66(3) : 347-358.

Unger, Jonathan, Collective inventives in the Chinese countryside : Lessons from Chen village," *World Development*, May 1978, 6(5) : 583-601.

Wong, John, Some aspects of China's agricultural development experience : Implications for developing countries in Asia," *World Development*, June 1976, 4(6) : 485-497.

Wood, Geof, "Rural development and the post-colonial state : administration and the peasantry in the Kosi Region of North-East Bihar, India," *Development and Change*. July 1977, 8(3) : 307-323.

INDEX